Process Management and Organizational Process Maturity

Anna Kosieradzka · Katarzyna Rostek

Process Management and Organizational Process Maturity

Economic and Non-Economic Organizations

palgrave
macmillan

Anna Kosieradzka
Faculty of Management
Warsaw University of Technology
Warsaw, Poland

Katarzyna Rostek
Faculty of Management
Warsaw University of Technology
Warsaw, Poland

ISBN 978-3-030-66799-3 ISBN 978-3-030-66800-6 (eBook)
https://doi.org/10.1007/978-3-030-66800-6

This Palgrave Macmillan imprint is published by the registered company Springer Nature Switzerland AG
The registered company address is: Gewerbestrasse 11, 6330 Cham, Switzerland

Introduction

The process approach is a dynamic management strategy, supplemental to the static method (functional and systemic). It entails such activities as: identification of processes, determination of their respective priorities and correlations, implementation of quality assurance and evaluation methods, performance monitoring, control, verification, and improvement. The effects of the process approach and its ultimate results are determined by the organization's capacity to ensure the proper implementation of each of the aforementioned elements that comprise the process lifecycle. At the same time, a process that is not subject to systematic control and evaluation, and consequently not adapted to the dynamically changing circumstances and requirements, is bound to quickly fail to serve its function at the expected level. This is a key challenge faced by processually managed organizations.

Although international studies conducted by the Business Process Trends organization between 2005 and 2017 indicate that over half of worldwide organizations are engaged in large-scale and strategically vital process management projects (BPTrends Report 2018, p. 9), the results of local research suggest significantly lower indices. As follows from a study conducted by ICAN Research for Macrologic SA[1] in 2016, only one in five Polish enterprises elected to implement the process approach. In turn, a study by Webcon and Computerworld conducted in 2017 indicated that only 30% of directors in Polish companies employing more than 150 people declared having no immediate needs in terms of process management. At the same time, the remaining 70% identified the following challenges and barriers in terms of the same (Wróbel 2017): personnel and competence shortages, lack of support from managerial staff for the process-centered perception of the organization, necessity to continuously adapt processes to ongoing changes in the organization itself and its milieu, difficulty in evidencing the profitability of investment in process reorganization (e.g., in terms of the ROI index), extensive and strong hierarchical dependencies in the context of knowledge, information, and decision-making flows (so-called information silos), high structural complexity of respective processes and difficulties in terms of their precise identification.

Given the above and the authors' own experience, one can assume that effective implementation of the process approach will be conditional upon a variety of factors depending on the character of the organization, its structure, number of employees, type of activity, as well as certain social, economic, and legal circumstances specific to the same. Hence, the methods, techniques, and tools employed in the context of process implementation and exploitation ought to account for and be adapted to said circumstances, the most important of which seems to be the type of the organization itself. For this reason, this monographic study will provide a comparative presentation of both economic and non-economic organizations with the aim of answering the following research questions:

Q1: To what extent can one identify differences in the process approach applicable to economic and non-economic organizations?

and

Q2: What is the importance of accounting for said differences relative to the efficiency and effectiveness of process results?

[1] http://organizacjahoryzontalna.pl/procesowosc-to-sie-oplaca/ <read 2019-10-18>.

Given the rather general character of the concepts of an economic[2] and non-economic[3] organization relative to our ability to viably approach the aforementioned questions, the perspective of these considerations had to be somewhat narrowed. Therefore, the characteristics provided in the subsequent chapters will focus specifically on production companies, as an example of economic organizations, and public administration entities, to represent the group of non-economic organizations. As a result, the discussion on economic organizations will study cases of processes related to production, manufacturing, and production services,[4] while in order to characterize non-economic organizations, the main focus will be placed on primary processes[5] and public services.[6] This narrowing of the scope will not hinder the main goal of the monographic study, i.e., identification of organizational differences and the significance of their impact on the efficiency and effectiveness of the process approach implementation in organizations. At the same time, we believe that the reader should benefit from the fact that the presented examples and case studies will be drawn from specific contexts limited as described above.

In order to comprehensively approach the adopted research questions, the structure of this monographic study will reflect an adapted version of Demin's four-stage PDCA cycle which includes all key phases of the process lifecycle in an organization, from its initial modeling to eventual improvement, and facilitates ongoing updates and adjustments vis-à-vis the continuously evolving needs of the organization (Fig. 1). A more detailed explanation of why this particular set of stages and process life-cycle structures are adopted will be provided in Chapter 2.2 below.

[2]Economic organization—an entity conducting economic activity aimed at producing goods and/or rendering services to generate profit.

[3]Non-economic organization—an entity conducting activity other than that defined as economic, e.g., provision of public services.

[4]Production services—material services directly satisfying specific human needs, e.g., installation and assembly robots, manufacture of products on commission, services related to the sphere of product manufacture, power distribution, etc.

[5]Primary processes—processes executed by public administration entities as part of their primary activities. Administrative processes aim to prepare and render public services to citizens/customers.

[6]Public services—services rendered by the public administration to citizens directly (as part of the primary activity) or indirectly by financing private entities rendering said services (by way of outsourcing).

At this stage, it should only be mentioned that the proposed scheme constitutes the summary and generalization of analyses conducted on a variety of different process lifecycles in organizations, and is based on the assumption that, on the one hand, all key areas of process management need to be accounted for, and on the other, that the level of generalization must ensure its applicability to any type of organizations.

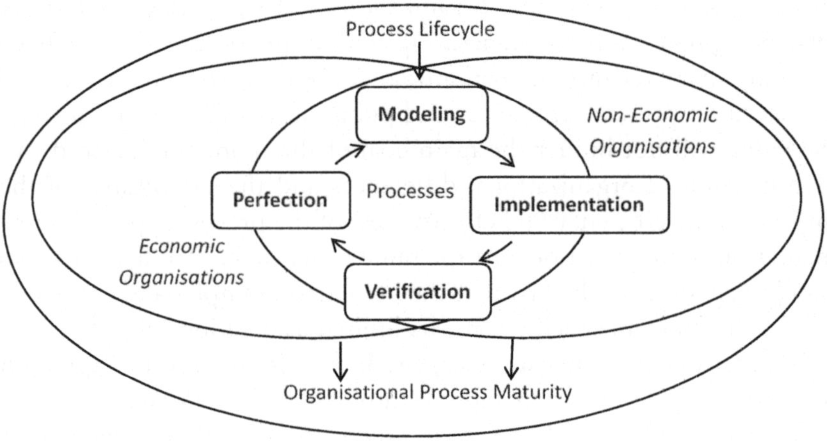

Fig. 1 Abstract idea *(Source: own elaboration)*

Due to the fact that the obtained results are affected by the organization's process maturity, a chapter related thereto is also included in the study. The proposed structure facilitated a broad elaboration on the problem—with insights related to the entire organization and its presence on the market as well as its overall business milieu, rather than limiting oneself to solely the internal perspective.

In light of the research results mentioned in the introduction, it should be concluded that the key barrier to the correct and comprehensive implementation of process management in organizations is posed, on the one hand, by the lack of competence and understanding on the part of managerial staff, and on the other, by the complexity and multidimensionality of the issue. Recognizing those needs, this monographic study presents and organizes concepts and problems in the area of process management, as well as indicates corresponding support methods, techniques, and tools in each of the four primary stages in the process lifecycle (i.e., modeling, implementation, verification, and perfection). Following the adapted structure of the PDCA cycle, the study identifies the possibilities, conditions, and effects that may be generated at each of the respective stages. At the same time, by considering production companies and public administration institutions separately, we were able to ensure adequate flexibility of the approach and possibility of applying the same to diverse types of organizations. The study concludes with a presentation of selected models of organizational process maturity as a tool facilitating objective and comparable measurement of process development and perfection relative to a variety of goals and expected results.

It is from the comparison of the two perspectives—economic and non-economic, as well as the comprehensive insight into process management viewed from the angle of improvement and development of an organization as a whole, that the primary value of the monographic study is derived. This distinction proves even more significant should one consider the scope of reasons that motivate organizations worldwide to make the effort necessary to implement and develop the process approach (BPTrends Report 2018, p. 10) with a view to facilitating improvement in terms of: operating efficiency and productivity, ability to flexibly respond to change, perception of the organization by customers, operating coordination and control, compliance, or efficient reorganization. When considering only two primary types of economic and non-economic organizations, it is clear that the specification of the aforementioned reasons will be of entirely different significance and will have to be done in a way dedicated to each of the two types, respectively.

The presented solutions and methods were largely developed in the course of studies and R&D projects in which the authors were involved over the course of a dozen or so years. Its value stems from the ability to demonstrate the manner and scope to which the type of the organization should impact the adopted approach to process implementation, but also recommendations for methodologies and tools applicable to the given type. The monographic study is therefore valuable on practical, theoretical, and methodological levels. It may facilitate the processes of education and training addressed to both students and practitioners but can also serve as a source of inspiration for researchers exploring issues of processual and organizational excellence.

Anna Kosieradzka
Katarzyna Rostek

Bibliography

BPTrends Report. (2018). https://www.redhat.com/cms/managed-files/mi-bptrends-state-of-bpm—2018-survey-analyst-paper—201803-en.pdf. Read 22 July 2020.

Wróbel, Ł. (2017). *Zarządzanie zmianą i procesami biznesowymi w polskich przedsiębiorstwach*. Raport z badania. Webcon, Computerworld.

Contents

List of Figures

List of Tables

1

The Multifaceted Character of Process Management in Organizations

Anna Kosieradzka

1.1 Classifications of Organizational Processes—Anna Kosieradzka

1.1.1 A Process—Review of Definitions

The concept of a process dates back to the classical school of management developed in the early years of the twentieth century by people engaged in the study and development of methodologies aimed at improving productivity, including Frederic Taylor, Frank and Lilian Gilbreth, Henry Gantt, Harrington Emerson, and Henry Ford. The latter made a particularly significant contribution to furthering the development of process management. In 1913, Ford opened the first vehicle production line which allowed him to overcome the difficulties inherent in craft manufacturing and ensure uninhibited flow of products as well as repeatability of the production process (Womack et al. 1990, p. 27).

It is hardly surprising that process-oriented thinking is rooted in production as this context clearly evidences how the manufacture of each product is the direct result of executing a sequence of activities described

A. Kosieradzka and K. Rostek, *Process Management and Organizational Process Maturity*,
https://doi.org/10.1007/978-3-030-66800-6_1

as the production process. It is noteworthy that the second faction of the classical school of management, the so-called administrative approach, did not identify processes at the time, focusing instead on managing the organization as a whole.

The concept of business processes covers all the activities executed in an organization, starting from the definition of customer needs and requirements, and ending with the supply of goods/services meeting said requirements ("end-to-end" approach).

Table 1.1 presents example definitions of a process as proposed by leading authors researching the process approach to management and the BPR concept.

As we can see, the key elements reemerging in the respective definitions include:

- chain / sequence of activities,
- output (product or service),

Table 1.1 Definitions of process

Author(s)	Definition
H. J. Harrington (1991, p. 9)	Any activity or group activities that takes an input, adds value to it, and provides an output to an internal and external customer
H. Davenport (1993, p. 7)	Process is simply a structured, measured set of activities designed to produce a specified output for a particular customer or market
M. Hammer and J. Champy (1993, p. 39)	Process as a collection of activities that takes one or more kinds of input and creates an output that is of value to the customer
M. Hammer (2010, p. 4)	[...] definition of process: end-to-end work across an enterprise that creates customer value
J. Oakland (2014, p. 199)	Organizations create value by delivering their goods and/or services to customers. Everything they do in that whole chain of events is a process.

Source Own elaboration

- customer (internal or external),
- generating value for the customer.

The 1990s saw a significant revival of the process approach and its viability in all types of organizations in the context of the dynamically developing information technologies and the ongoing globalization of the economy, which bore fruit in the form of one of the key concepts in contemporary management—BPR. The authors of one of the first books exploring the problems of BPR, M. Hammer, J. Champy defined Business Process Reengineering as "the fundamental rethinking and radical reengineering of business processes to achieve dramatic improvements in critical, contemporary measures of performance, such as cost, quality, service and speed" (1993, p. 51). The definition continues to be widely accepted as it encapsulates the gist of process excellence and its impact on organizational performance.

1.1.2 Classification of Processes

The wealth of literature pertaining to the process approach offers sometimes quite diverse process classifications. Presented below are the ones that had the most significant impact on the evolution of the process approach.

After M. Porter

The value chain concept, developed by M. Porter for the purposes of strategic analyses in business, provided the foundation for the modern the process approach. The value chain is a sequence of actions allowing a company operating within a given sector to launch a valuable output (product or service).

According to Porter (1985, pp. 11–15), most organizations engage in hundreds, even thousands, of activities in the process of converting inputs into outputs. These activities can be classified generally as either primary or support activities.

The primary activities are essential in adding value and securing competitive advantage, and include:

- inbound logistics:—arranging the inbound movement of materials, parts, and/or finished inventory from suppliers to manufacturing or assembly plants, warehouses, or retail stores,
- operations—all the activities required to transform inputs into outputs (goods and services),
- outbound logistics—all the activities required to collect, store, and distribute the output to the end user,
- marketing and sales—activities aimed at informing buyers about goods and services, inducing buyers to purchase them, and facilitating that purchase,
- service—all the activities required to keep the product or service working effectively for the buyer after it is sold and delivered.

Support activities help make primary activities more effective and include:

- procurement—the acquisition of goods, services or works from external sources,
- human resources management—all activities involved in recruiting, hiring, training, developing, compensating and (if necessary) dismissing or laying off personnel,
- technological development—pertains to the equipment, hardware, software, procedures and technical knowledge brought to bear in the firm's transformation of inputs into outputs,
- infrastructure: activities such as accounting, legal, finance, control, public relations, quality assurance and general (strategic) management.

Porter was the first to observe that the process of generating value for the customer runs across the entire function of the enterprise, and that the effectiveness of an organization depends on the performance levels achieved at the meeting points between the respective links in the value chain, thus formulating the basis for the development of process management. While primary processes were fairly conclusively identified, the weakness of his value chain stemmed from the somewhat less precise definition of the criteria determining the identification of auxiliary (support) activities. This shortcoming was subsequently eliminated in the classification proposed by Rummler and Brache, who singled out management processes from the group of support activities.

After G. Rummler and A. Brache

Rummler and Brache (1995, pp. 75–77) identified three key groups of processes:

- Primary processes are those leading to the creation of an output (product or service) launched on the market; general primary processes include for instance: the establishment of a new company, development and launch of a new output (product/service), production, distribution, order fulfilment, invoicing, and warranty service; primary processes specific to respective industries include e.g.: processing loan applications (banking), processing insurance claims (insurance), awarding subsidies (governmental agencies), product returns (retail), preparation of meals (restaurants), luggage transport (airlines), subscriber services (telecommunications), printing (printing houses), or booking (hotels);
- Support processes are those whose outputs remain invisible to external customers, but that are vital to the correct implementation of primary processes, e.g.: plan distribution and production records, recruitment, training, product control, procurement, IT management;
- Management processes include all activities aimed at ensuring operating efficiency and effectiveness of the organization, e.g. strategic and operations planning, identification of goals, budgeting, asset allocation, employee assessment, monitoring and evaluation of operations.

Rummler and Brache, who identified three distinct levels of productivity, namely: organizational, processual, and workstation-related, believed processes to be the key determinants of effectiveness and emphasized the fact that "*an organization is only as effective as its processes*" (Rummler and Brache 1995, p. 76).

In the Perspective of Value Creation

The concept of Lean Management introduced the notion of the Value Stream which encompasses three types of activities (Womack and Jones 1996, pp. 19–20):

- Value-adding, related to generating the value of an output for customers and serving the satisfaction of their requirements; these are activities for which customers are willing to pay as they increase their satisfaction or directly solve their problems;
- Operating (efforts, activities), not adding value but facilitating its generation and unavoidable given the currently available technologies and resources (NVA—*nonvalue-adding but necessary*). These typically include activities related to control, transport, administration, reporting, etc. In Japanese, such activities are referred to as *muda* (Eng. waste)—type I losses whose reduction or elimination requires technological, organizational, or resource-related changes;
- Waste—operations that do not add value and are superfluous, liable for immediate elimination, e.g. lead time, overlaps, fixing errors made at earlier stages, etc. Such activities are commonly referred to as type II *muda*. Usually, they are not related to satisfying customer needs but rather solving problems within the organization and as such, they are considered wasteful.

In the authors' opinion, this classification is the most important from the perspective of process excellence. Every action that does not contribute to generating value but instead leads to futile personnel engagement should be considered a loss. Process improvement requires one to identify nonvalue-adding and wasteful processes (type I and II *muda*) and to implement solutions aimed at eliminating or at the very least minimizing the same.

APQC Process Classification Framework

The American Productivity and Quality Center (APQC) developed the most comprehensive process classification framework. Their PCF is a hierarchical list of business processes. It includes 13 high-level categories of work (Fig. 1.1).

These 13 Categories (Level 1) break down into increasingly granular levels:

- Level 2—Process group: a group of processes that are part of executing a category,
- Level 3—Process: a single process,
- Level 4—Activity: a key step performed to execute a process,
- Level 5—Task: an element of work that goes into executing an activity.

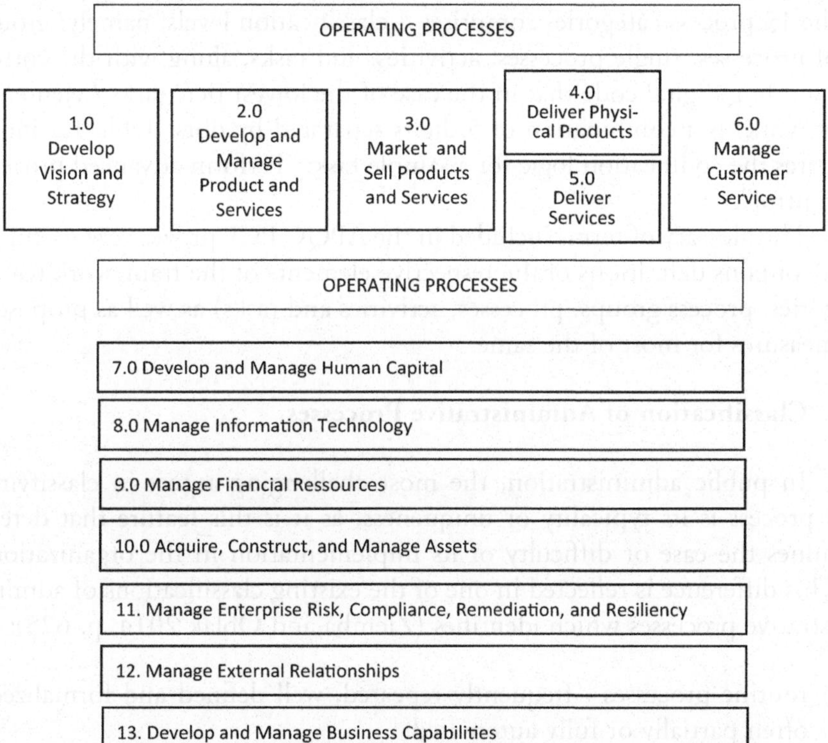

Fig. 1.1 APQC Process Classification Framework (PCF) (*Source* https://www.apqc.org/ [14 July 2020])

Table 1.2 Example task coding under the APQC PCF

4,0 Deliver Physical Products	4				
Produce/Assemble/Test product	4	3			
Produce/Assemble product	4	3	2		
Monitor and optimize production process	4	3	2	5	
Perform advanced process control	**4**	**3**	**2**	**5**	**2**

Source Own elaboration

The Process Classification Framework can be a useful tool in understanding and mapping business processes. (Oakland 2014, p. 205).

The framework is continuously developed and perfected. The current version of the APQC PCF is available at https://www.apqc.org/. Each of the 13 process categories comprises 4 classification levels, namely: group of processes, single processes, activities, and tasks, along with the corresponding digital code that in the case of the lowest tier (tasks / elements of work) is a combination of 5 digits separated by dots. Table 1.2 illustrates the codification logic for a sample task: "Perform advanced process control".

The glossary of terms included in the APQC PCF proves very useful as it contains definitions of the respective elements of the framework (categories, process groups, processes, activities and tasks) as well as proposed measures for most of the same.

Classification of Administrative Processes

In public administration, the most challenging aspect in classifying a process is its typicality or uniqueness, as it is this feature that determines the ease or difficulty of its implementation in the organization. This difference is reflected in one of the existing classifications of administrative processes which identifies (Ziemba and Oblak 2014, p. 625):

- routine processes—frequently repeated, well defined and formalized, often partially or fully automated,
- case-by-case processes—situations where a standard process cannot be implemented due to certain unique circumstances that require the case to be considered in a broader context (legal, social, personal),

- processes requiring negotiation—characterized by the involvement of several parties, which necessitates an in-depth analysis of the situation and all factors relevant to the eventual decision,
- poorly structured processes—lacking assigned procedures or defined standards (e.g. sporadic events), where each instance requires a customized approach.

Even a preliminary analysis of the above classification allows the diagnosis of the challenge that will have to be faced by an organization if it is to shift from functional to process management.

1.2 Process Management—Anna Kosieradzka

The process approach is rooted in the general systems theory first proposed by L. von Bertalanffy in the 1930s. It was later adapted by G. Nadler (1967) to the needs of effective organization in a *work system* comprising a combination of *human, physical, and financial resources* that allow organizations to convert *inputs* (materials, information, human labor) into viable *outputs* (goods or services). The function of this work system is described as the *purpose* that the system serves, while the sequence of actions necessary to convert inputs into outputs is a *process*. The physical, economic, and social factors that shape the milieu wherein the work system functions is described as the *environment*. The correlations between said elements are presented in Fig. 1.2.

Nadler's IDEALS may be considered as an inspiration for the concept of Business Process Reengineering as it emphasized the need to comprehensively organize an effective work system by defining the following elements: purpose, processes, inputs, outputs, resources, and the environment.

1.2.1 Business Process Orientation

Over the years, various orientations dominated the development of scientific management, e.g.: market-centric, strategic, result, human,

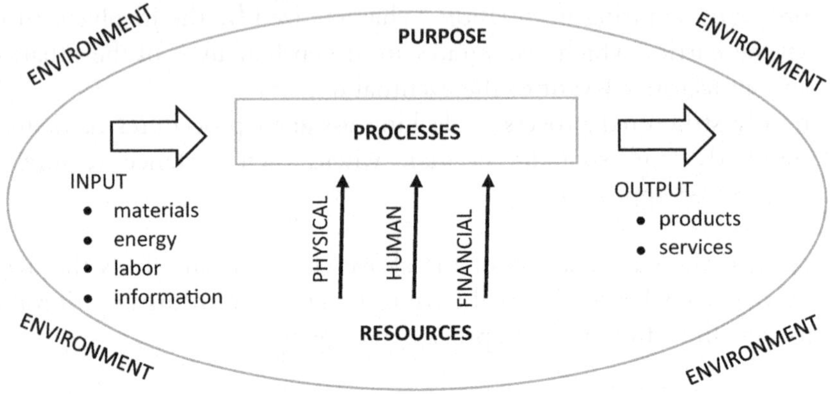

Fig. 1.2 Elements of the **system** concept (*Source* own elaboration)

quality-centric, process, shifts, knowledge, etc. depending on the primary goals adopted by top tier management.

The Business Process Orientation (BPO) also known as the process approach is currently considered to be among the most important ideas pertinent to the management of contemporary organizations, be it production companies, service providers, or administrative organizations. One should note at this point the interchangeable use of the terms process orientation and the process approach in literature, both denoting a process-oriented approach to management.

But what exactly does "process-oriented management" mean? Rummler and Brache, (1995, pp. 32–33) identified a phenomenon characteristic of traditional management which they referred to as the "silo effect". Such silos (i.e. tall, thick-walled buildings deprived of windows) are formed around certain functional departments and hinder, or even downright prevent at lower tiers of the organization, solving problems shared simultaneously by several departments. The silo effect leads to a situation where top tier managers are forced to solve problems emerging at lower levels, which distracts them from more important issues related to customers or the competition. Heads of respective departments can sometimes be so conflicted that inter-functional problems are simply not discussed at all. Striving to achieve the best possible results within respective departments can be detrimental to the performance of the company as a whole. The silo effect does not affect processually managed organizations where the greatest possibilities for improvement are actually found at the meeting points between the departments. Table 1.3 presents the key differences between functional and process-oriented organizations.

As posited by McCormack (2001), the basis for BPO consists of the following:

- process view—documentation of "end-to-end" processes,

Table 1.3 Characteristics of functional and process-oriented organizations

Functional organization	Process-oriented organization
Focus on the internal policies and goals of functional departments	Focus on customer needs at all stages of the process
Division of labor and focus on the assigned tasks	Introduction of the "vendor-customer" relationship, differentiation between internal and external customers. Focus on the process as a whole
Focus on the efficient performance of respective tasks	Focus on value generation. Potential for optimization, reduction of losses and nonvalue-adding activities.
Elimination of the consequences of variability. Maintaining reserves to compensate for disturbances	Elimination of the root causes of process variability
Scopes of responsibility distributed among functional managers	Clear scopes of responsibility (process owners)
Hierarchical management structures, centralized authority that prolongs the decision-making process	Flat management structures, delegation of authority
Individualism, both in terms of departments and staff evaluations	Teamwork, empowerment, evaluation of teams rather than individual staff members
Teams competing for resources	Collaboration rather than competition between respective teams
Narrow employee specialization, unwillingness to take on tasks that require additional competences	More comprehensive tasks assigned to respective employees, which necessitates development of human resources
Authority limited to a specific function, hindered inter-functional communication (silos)	Efficient communication and knowledge management

Source Own elaboration

- process tasks—the identification of a process owner rather than functional manager,
- process management and measurement—components that measure output quality, cycle time, process cost and variability.

Nadarajah et al. (2016) expand this basic logic by adding "values and beliefs" and defining BPO as "cross-functional and outcome-oriented with emphasis on business processes, jobs and structures, management and measurement systems and finally value and beliefs".

Tables 1.3 illustrate the reasons why process orientation is so important to contemporary organizations.

Without understanding how respective processes actually operate, one cannot hope to improve the functional productivity and efficiency of the organization. The process approach facilitates the improvement of the entire organization's flexibility and responsiveness to change. The same pertains not only to particular processes analyzed individually, but also to the mutual correlations between the same.

Even though the process approach has been intensively developed for over 20 years, we still observe its less than sufficient implementation in organizations all over the world. The key elements comprising the process approach are well known, however, it is the correct combination and incorporation of the same in an organization that poses a challenge and is the focus of studies conducted within the dedicated subdiscipline of Business Process Management (BPM).

1.2.2 Business Process Management

Literature provides numerous definitions of Business Process Management. Lee and Dale (1998) reviewed interpretations of BPM posited in texts published in the 1990s and identified key elements that can be said to comprise the classical basis of the same:

- horizontal organizational structure,
- documented processes,
- structured and standardized operational activities,

- availability of systems and procedures to ensure consistency of performance,
- implementation of performance tracking and measurement,
- continuous improvement through problem solving; based on the scientific approach.

A popular and nowadays commonly accepted definition, also included in Wikipedia, was proposed by Jeston and Nelis (2014, pp. 4–5): "BPM is a discipline in operations management in which people use various methods to discover, model, analyze, measure, improve, optimize, and automate business processes".

Clearly, the concept of Business Process Management has evolved considerably over the last century, from one rooted in the reality of classical production-centered school of management popular in the early twentieth century, through the process orientation of the 1990s, to contemporary integrated process management combining BPM with project management and knowledge management. A. Bitkowska (2019, pp. 18–19) identified 5 stages in the development of process management, as presented in Table 1.4.

The first stage in the development of the process approach can be traced back to the 1st half of the twentieth century when it focused on improving the efficiency of production processes. Eminent representatives of this school, later referred to as Industrial Engineering, included F. W. Taylor, L. Gilbreth, F. Gilbreth, H. Ford, H. Emerson, H. L. Le Châtelier, H. L. Gantt, Karol Adamiecki, and E. Hauswald. Research was conducted within the scope of so-called Time and Motion Study, which yielded such methods as continuous observation and work sampling, process flow charts and material flow charts, worktime normalization, etc., which aimed to identify work and break times with a view to improving process productivity. A number of methods of timewise workflow organization were also proposed in that period: K. Adamiecki's schedules—1898 or Gantt's charts—1910. Most of these methods are still in use today, naturally, nowadays employing IT solutions.

The second stage in the development of process management started with the emergence of the systemic approach in management sciences in the 1960s, and continued well into the 1980s. Eminent representatives

Table 1.4 Development of process management—BPM

Stage	Characteristics	Concepts, methods and techniques
Stage 1 Classical School of Management Time and Motion Study Industrial Engineering	Seeking ways to improve work productivity Work normalization Planning work Employing statistical methods to control the performance and results of production processes (Shewhart's control charts)	Work specialization, Normalization of work and analysis of work movements (Therbligi) Operations Research Process scheduling (Gantt's chart) Continuous and sampling observation (to identify operations generating value and loses) Statistical quality control Statistical process control (SPS)
Stage 2 Process approach	Systemic approach Value chain analysis Computerized production planning systems	Quality management concepts TQM, FMEA, QFD, MRP class systems TOC—Theory of Constraints TPS—Toyota Production System JIT—Just-in-Time ABC—Activity Based Costing

Stage	Characteristics	Concepts, methods and techniques
Stage 3 Process Management	Radical process restructuring Continuous process improvement Performance Management	Lean Management BPR Kaizen Benchmarking BSC—Balanced Score Card Process mapping (VSM—Value-Stream Mapping, BPMN—Business Process Modeling Notation)
Stage 4 Dynamic Process Management	Expansion of process management to include supply chains Continuous process improvement with employee involvement Generation and management of organizational and process knowledge	SCM—Supply Chain Management Six Sigma Intelligent organization and learning organization Models of organizational process maturity
Stage 5 Integrated Process Management	Integration, cohesion and flexibility of processes Integration of management concepts and methods rooted in process management Integration of process management and project management in the context of organizational knowledge	Knowledge Management Project Management Risk and Business Continuity Management Agile Business Process Management

Source Own elaboration after (Bitkowska 2019)

of this school included L. von Bertalanfy and G. Nadler, who defined an organization as a system, developed a viable conceptual toolset (See Fig. 1.2), and defined the notions of process and its context. At roughly the same time, the TPS—Toyota Production System was introduced in Japanese industry with a view to improving production efficiency by eliminating wastage—it would later become the prototype for Lean Management. TPS is a combination of a variety of techniques, concepts and principles aimed at eliminating the so-called 3M (i.e. *muri*—overload, handicap; *mura*—irregularity, variability; and *muda*—waste) and developed by Japanese authors, primarily T. Ohno and S. Shingo. Apart from strictly production-related instruments (jidoka, heijunka, kanban, andon, SMED) dedicated to the specific needs of TPS, the system also takes advantage of solutions developed under the TQM concept, quality management systems, as well as the Kaizen concept introduced into the scope of management by M. Imai. Other major developments at this stage included the definition of concepts crucial to process management, namely the value chain (M. Porter) and value stream (T. Ohno and S. Shingo).

The third stage in the development of BPM took place in the 1990s. The decade saw a true explosion of papers discussing BPM, with some of the major contributions including works by M. Hammer and J. Champy, who introduced the concept of Business Process Engineering into the realm of scientific management, emphasizing the significance of IT tools in the implementation of radical processual changes in organizations. Researchers also noticed the possibilities and potential derived from continuous improvement of business processes through evolutionary, rather than revolutionary changes, in accordance with the concepts of Lean Management, TQM, and Kaizen. An important insight made in this period was that the process approach could actually be applicable to a variety of areas such as project management, HR management, quality management, or risk management.

The fourth stage is often described as dynamic process management which, on the one hand, extends beyond the bounds of a single organization in an effort to optimize processes within entire supply chains or various networking structures established by organizations, and on the other, emphasizes better utilization of knowledge and human development. The aim is to achieve greater process flexibility to allow better adaptation of processes to the changing conditions in the environment or customer requirements. The primary concept of dynamic process management is referred to as Agile Management. However, while Lean Management offers a wide range of methods and techniques to facilitate its practical implementation, the level of operationalization observed for Agile Manufacturing is somewhat low. Improvement of processes is done with a view to reducing their variability with the use of statistical methods derived from the Six Sigma toolset.

The fifth stage of integrated process management is currently still in the early phase of its development. Integration is to affect areas of management that have, until now, been perceived separately: Business Process Management and Project Management. In most organizations, process management and the management of strategic project portfolios need to be integrated due to their reliance on the same resources: human, technical, financial, and information, as well as the same knowledge base generated in the organization. This is the most recent trend observed in the context of business process management.

As follows from the above historical overview, process management originated from the management of production processes which, as befits the very nature of the same, constitute methods of executing technological processes designed to yield specific products. The search for ways of improving the efficiency of production processes inspired the development of many concepts, methods, and techniques later adopted into Business Process Management and successfully employed also by service providers and administrative organizations.

1.2.3 Methods and Concepts Developed on the Base of the Process Approach

The process approach has found a permanent place in contemporary management systems. As such, management concepts entail organization and generalization of practical experiences. Their structure can be divided into two primary layers: the general idea (philosophy) formulated as a list of general goals, assumptions, and principles on the one hand, and specific guidelines and tools (methods, techniques, procedures) necessary to practically implement the premise of that general idea on the other. Hence, a management concept can be defined as a cohesive and comprehensive approach to solving problems that emerge when managing an organization, complete with a set of rules and guidelines as well as an array of methods and tools (either dedicated to this particular concept or derived from others). Table 1.5 presents the most important contemporary management concepts utilizing the process approach.

These concepts are most commonly implemented in organizations that are independently or incidentally related to each other. They mutual interactions, as well as shared methods and techniques are not sufficiently exploited. However, the concepts are so strongly correlated that it is sometimes difficult to determine whether the given solution is implemented under TQM or LM. There have also been reports of certain concepts being combined, e.g. Lean Six Sigma (a combination of LM and Six Sigma—Lunau 2008, pp. 5–18), or TLS (a combination of TOC, LM, and Six Sigma—Cox and Schleier 2010, pp. 1067–1080).

A large array of methods and techniques are utilized under the respective concepts. For instance, elimination of losses and continuous improvement are notions shared by all of the same, SMED is a tool included in both LM and TOC, statistical process control (SPC) is mentioned as a functional element in TQM, Six Sigma, as well as Lean Manufacturing. Table 1.6 presents selected methods and tools utilized under various concepts, as well as the co-dependence and complementarity of the same.

As follows from the above, the process approach serves as a common platform for the implementation of various management concepts in organizations. Indeed, one could perceive said concepts as tools facilitating the introduction of the process approach.

Table 1.5 Contemporary management concepts utilizing the process approach

Concept	Goal—primary idea	Rules/steps
LEAN MANAGEMENT	A concept that perceives the use of resources for any purpose other than generating value for the customer as wasteful (Womack and Jones 1996)	(1) Value—Specify the value desired by the customer. (2) The Value Stream—Identify the value stream for each product (3) Flow—Make the product flow continuously through the remaining value-added steps (4) Pull—Introduce pull between all steps where continuous flow is possible (5) Perfection—Striving for perfection through continuous improvement
TQM	By adopting appropriate principles of management, organizations can increase quality and simultaneously reduce costs (by reducing waste rework, staff attrition and litigation while increasing customer loyalty). The key is to practice continual improvement and think of manufacturing as a system, not as bits and pieces (W. E. Deming)	In accordance with ISO 9000:2000: (1) Customer orientation, (2) Leadership, (3) Employee engagement, (4) **The process approach**, (5) Systemic approach to management, (6) Continuous improvement, (7) Fact-based decisions, (8) Mutually beneficial relations with suppliers
BUSINESS PROCESS REENGINEERING	Fundamental rethinking and radical reengineering of **processes** in the organization, leading to dramatic (game-changing) performance improvement (Hammer and Champy 1993)	(1) Selection of the process (processes) to be reengineered (2) Establishment of a team responsible for implementing the changes (3) Understanding the process (4) Reengineering the process (5) Implementation of the improved process

(continued)

Table 1.5 (continued)

Concept	Goal—primary idea	Rules/steps
KAIZEN	Continuous perfection of **processes** in the organization through minor improvements implemented by all the staff members (Imai 1986)	(1) Maintenance and improvement of standards—the maintenance relies on Kaizen activities, while improvement can be considered in terms of Kaizen efforts and innovations, (2) **Process orientation**—perfecting processes as the basis of performance improvement, (3) Application of Deming's PDCA and SDCA cycles, (4) Quality comes first—the most important goals of the enterprise concern quality, (5) Data operation—reliance on data pertinent to the current circumstances when attempting to solve a problem, (6) The next process is the customer—distinction between external customers (on the market) and internal customers (within the company), (7) Engagement of all employees—equally at managerial and front-line levels

Concept	Goal—primary idea	Rules/steps
THEORY OF CONSTRAINTS	Every organization has only a few constraints, elimination of which leads to radical improvement of results (Goldratt and Cox 1992)	Searching for improvements is a 5-step process: (1) Identifying the constraint (2) Exploiting the constraint (3) Subordinating all other resources and activities to the constraint (4) Elevating the constraint (5) Returning to step (1) to complete the cycle of continuous improvement
SIX SIGMA	Six Sigma is a complex and flexible system of achieving, maintaining and increasing success in business. It is characterized by understanding customer needs and using facts, data and statistical analysis results (Pande et al. 2001)	The improvement process based on the DMAIC cycle: (1) D—Defining projects, problems, measurements, reference levels, aspects critical to quality (2) M—Measuring the current state of the key **process**, establishing and verifying the process measurement system (3) A—Analysis of statistical data allowing the most important factors affecting the defined critical aspect to be identified (4) I—Improvement in order to reduce the level of defects and deviations (5) C—Controlling aimed at maintaining the achieved quality level

Source Own elaboration based on (Kosieradzka et al. 2011)

Table 1.6 Selected methods and tools utilized under the respective management concepts

Methods and tools	LM	Kaizen	TOC	TQM	Six Sigma	BPR
Process approach	+	+	+	+	+	+
Customer Value						
Continuous improvement	+	+	+	+	+	+
Process mapping	+	+	+	+	+	+
Value stream, value chain	+	+	+	+	+	+
Waste elimination	+	+	+	+	+	+
JIT	+	+				
Pull system/Kanban	+	+	+	+		
Standardization	+	+	+	+	+	+
5S	+	+	+	+		
Poka-Yoke	+	+	+	+	+	+
Visual control	+	+	+	+	+	+
SPC	+	+		+	+	
PDCA/SDCA (Deming cycle)	+	+	+	+	+	+
SMED	+	+	+			
FMEA		+		+	+	
Statistical methods and tools	+			+	+	
Teamwork and human resources involvement	+	+		+	+	

Source Own elaboration based on (Kosieradzka et al. 2011)

1.3 Needs and Challenges in Terms of Process Management—Anna Kosieradzka

1.3.1 Business Process Management vs. Process Management

Process management can be analyzed from two perspectives:

- A broad view understood as a holistic, comprehensive approach to the management of a modern organization whose scope encompasses the levels of both strategic planning and operations, and which entails the systematic application of respective concepts, methods, and tools to influence processes performed in accordance with the organization's strategic goals. The perspective relates to the entire institution in all organizational, social, financial, IT and knowledge related aspects of the same (Bitkowska 2019, p. 25);

- A narrow view, focused on the level of operations and understood as the utilization of various methods with the aim of identifying, modeling, analyzing, measuring, improving, and optimizing business processes (Jeston and Nelis 2014, p. 4).

It is noteworthy that the two perspectives are closely intertwined as the improvement of processes in an organization should, by principle, aim to achieve certain strategic goals. However, simply managing respective processes is not synonymous with the introduction of process management in the organization as a whole. Hence, management of processes that entails identification, modeling, analysis, measurement, improvement, and optimalization of the same is necessary but not in itself sufficient to talk of process-oriented management. Apart from management of processes as such, organization-wide process-oriented management must also account for the following considerations:

- The organization's goals and strategy should be process-oriented,
- The process architecture should be consistent, i.e. the respective correlations between internal processes as well as between processes and the environment should be cohesive,
- The organization's technical infrastructure (technical resources, IT systems) and social milieu (organizational structure, incentive system, training) ought to be process-oriented,
- The human resources (staff members and managers) ought to understand the processes, be enterprising and creative, willing to analyze and improve as needed.
- The development of this approach is facilitated by training and incentive systems rewarding proactive attitudes,
- The organization's knowledge resources, including methods and techniques employed in the management of processes, measures and data needed for their definition, as well as systems used in their monitoring should support process management.

In recent years, the importance of utilizing organizational knowledge as one of the key elements facilitating process management has been noted. In their paper, Linderman et al. (2010) described a Process Management System Knowledge Framework, see Fig. 1.3.

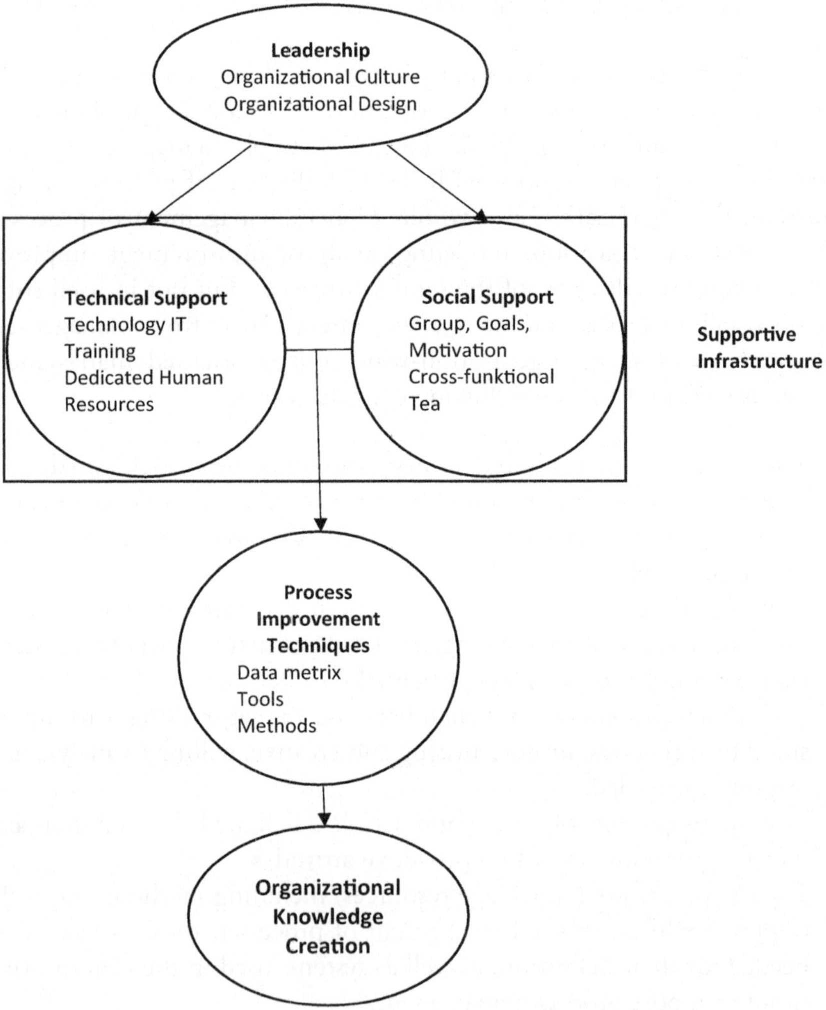

Fig. 1.3 Process management system knowledge framework (*Source* Linderman et al. 2010)

With this approach, a specific knowledge culture is created within the organization that entails standardized methods of gathering, sharing, and converting knowledge to become more accessible and available via numerous channels. Management of organizational knowledge is hardly a new trend in management, however, it was not until the last decade that papers would surface emphasizing the need to integrate process management with knowledge management (see Table 1.4, stage 5) with a view to improving the effectiveness of BPM implementation.

1.3.2 Key Success Factors and Limitations in the Implementation of Process Management

Rosemann and vom Brocke (2015) identified the following six key factors as elements of successful process management:

- Strategic Alignment—which means that BPM has to be adapted to the general strategy of the organization. Processes should be designed, executed, managed, and measured with due consideration for strategic priorities and the specificity of the strategic situation;
- Governance—that defines relations with stakeholders and ensures appropriate transparent liability in terms of roles and responsibilities applicable to the respective levels of BPM (portfolio, program, processes and operations);
- Methods are defined in the context of BPM and a set of instruments and techniques facilitating activities across the entire life-cycle of a process and in the context of BPM initiatives introduced throughout the organization, for instance process modeling, analysis, and improvement techniques;
- Information Technology determines the selection of IT solutions that are of vital importance in the context of BPM. The same include in particular process analysis (e.g. SPC—statistical process control) and process modeling support tools, but increasingly often also include support for everyday process execution (e.g. banking or insurance applications);

- People (individuals and groups) are key elements of BPM as they continuously increase their competences in terms of process management and ongoing process perfection with a view to improving business performance;
- Culture entails the establishment of an environment conducive to the organization's functioning. It is a set of collectively shared values and beliefs with regard to a process-centered organization.

For each of the aforementioned core elements, five process capability areas were further identified (Table 1.7).

Experience shows that organizations which make the effort to implement process management tend to be disappointed by the lower than expected efficiency and effectiveness of the solution. It seems that the reasons for this may lie in incompatibility between processes as such and ill-adjustment of processes to the various requirements and functional capacities of the organization.

The classification of key elements of process management presented in Table 1.7 became the starting point for the identification and study of factors potentially responsible for the inconsistency of processes in production companies (Jurczuk 2018). A. Jurczuk defined the concept of business process inconsistency as an obstacle or discrepancy in terms of the execution, understanding or management observed in the context of business processes, whose consequences need to be eliminated or minimized.

This holistic approach to the key elements determining the success of BPM implementation shows the multi-faceted character of problems related to the introduction of process management in organizations viewed broadly as a certain environment wherein all the organizational resources at the levels of both strategy and operations are targeted towards business processes with the aim of satisfying customer requirements and expectations.

In the report published by Capgemini (2012), 5 main internal barriers to BPM implementation were identified:

- Functional silo culture—55% of respondents identified functional silo culture as a barrier to process improvement at their organization,

Table 1.7 The Six Core elements of business process management

Factors	Capability Areas
Strategic Alignment	Process Improvement Planning Strategy & Process Capability Linkage Enterprise Process Architecture Process Measures Process Customers and Stakeholders
Governance	Process Management Decision Making Process Roles and Responsibilities Process Metrix and Performance Linkage Process Related Standards Process Management Compliance
Methods	Process Design and Modeling Process Implementation and Execution Process Monitoring and Control Process Improvement and Innovation Process Program and Project Management
Information technology	Process Design and Modeling Process Implementation and Execution Process Monitoring and Control Process Improvement and Innovation Process Program and Project Management
People	Process Skills and Expertise Process Management Knowledge Process Execution Process Collaboration Process Management Leaders
Culture	Responsiveness to Process Change Process Value and Beliefs Process Attitudes and Behaviors Leadership Attention to Process Process Management Social Networks

Source After (Rosemann and vom Brocke 2015)

making it the most frequently cited obstacle. Different parts of the organization tend to operate in functional silos,

- Fragmented budget—is closely related to the functional organizational structure. This method of working promotes efficiency within a particular silo but undermines the possibilities for working collaboratively and seamlessly across an organization,

- Perception of BPM as an IT item—48% of respondents identified the perception of BPM as an IT item as a barrier to effective process improvement,
- Resistance to BPM from IT staff responsible for existing systems—44% of respondents found that resistance from staff responsible for existing systems was a barrier to successful process improvement. IT departments will naturally be protective of current tools and argue that these can achieve value,
- Lack of readiness or willingness to change—46% of respondents identified a lack of change readiness or change willingness at their organization as a barrier to BPM adoption.

However, researchers studying Business Process Management from the theoretical perspective as well as practitioners of process management are now beginning to notice that further development of knowledge on process management and inclusion of new areas within its scope does not in fact lead to more effective BPM implementations in economic or non-economic organizations. For that reason, both researchers and practitioners are now considering ways to enhance the agility of business process management. On the one hand, Agile Process Management organizes structures and procedures with a view to facilitating repeatable business processes, and on the other, it ensures the possibility of immediately handling difficult cases that challenge the traditional, formalized techniques of process management. One can expect that this will be the direction in which Business Process Management is likely to develop.

The present book focuses on the management of processes in the narrow perspective defined above, with only Chapter 7, where models of process maturity in organizations are discussed, touching upon broadly understood business process management. The management of processes entails identification, modeling, analysis and measurement, improvement and optimization of business processes with the use of suitable methods and tools. The emphasis on continuous process improvement means that the aforementioned sequence of steps is largely repeatable. Hence, the subsequent sub-chapter focuses on the phenomenon of cyclicity observed in process management.

1.3.3 The Cycle of Process Management in Organizations

A key concept in the context of business process management is the so-called BPM lifecycle comprising repeatable stages, from identification of the problem, through formulation of a model, functional analysis and evaluation, to eventual process improvement. The stages and phases of the process management cycle in organizations have been described differently by various authors. Table 1.8 presents selected approaches to the problem of defining stages of the process management cycle.

There are evident similarities between the classifications proposed by the respective authors. At times, some stages are combined while others are omitted, which may reflect the given authors' experience or the specificity of the organizations they referred to when developing their models. For instance, Van der Aalst (2013) aligned the stages of the BPM lifecycle model with the model of process data management.

1.4 Chapter Summary—Anna Kosieradzka

The development of process management concepts has been ongoing for many years, nowadays it is considered to be a subdiscipline of management sciences combining both theoretical (definitions, classifications, principles, theoretical models) and practical considerations (experience and practical conclusions of managers, their requirements and expectations). The above chapter presented selected definitions of the concept of process as such, as well as process classifications developed from various perspectives, which will be referred to in the subsequent chapters.

The development of process management was presented, starting from the so-called classical school of management focusing on the improvement of production process efficiency, and ending with the most contemporary approaches integrating BPM—Business Process Management with 3PM—Project, Program, and Portfolio Management in the environment of an organizational culture facilitated by a knowledge management system. The key success factors as well as barriers and limitations to the implementation of process management were discussed

Table 1.8 Stages of Lifecycle Management

Source	Stages of process management lifecycle
Zur Muehlen and Ho (2006, p. 4566)	Goals specification, environmental analysis—organizational analysis Process design—specification and modeling Process implementation—Workflow modeling and implementation Process enactment—workflow execution Process monitoring—business activity monitoring Process evaluation—warehousing, controlling, process mining
Netjes et al. (2006, pp. 497–508)	Process design Process configuration Process execution Process control Process diagnosis
W. Van der Aalst (2013)	Process (re)design Process implementation and configuration Process execution and adjustment
Malinova and Mendeling (2018, pp. 882–889)	Process identification Process analysis Process reengineering Process implementation Process monitoring and control
Dumas et al. (2018, pp. 22–24)	Process Identification Process discovery (as-is modeling) Process analysis Process reengineering (improvement) Process implementation Process monitoring

Source Own elaboration

along with the differences between business process management (BPM) and management of processes as such. The differentiation between the two concepts is of crucial importance as this monographic study focuses on the management of organizational processes from the perspective of managing the process lifecycle.

The various approaches to the Process Management Lifecycle listed above provided a basis for the original organizational process management scheme presented in Chapter 2.

References

Bitkowska, A. (2019). *Od klasycznego do zintegrowanego zarządzania procesowego w organizacjach* [From Classical to Integrated Process Management in Organizations—Only in Polish]. Warszawa: C.H. Beck.

Capgemini. (2012). *Global Business Process Management Report*.

Cox, J. F., & Schleier, J. G. (Eds.). (2010). *Theory of Constraints Handbook*. New York: McGraw-Hill.

Davenport, T. H. (1993). *Process Innovation: Reengineering Work Through Information Technology*. Boston: Harward Business School Press.

Dumas, M., La Rosa, M., Mendling, J., & Reijers, H. A. (2018). *Fundamentals of Business Process Management* (2nd ed.). Berlin: Springer.

Goldratt, E., & Cox, J. (1992). *The Goal*. Great Barrington: The North River Press.

Hammer, M. (2010). *What Is Business Process Management?* In J. vom Brocke & M. Rosemann (Eds.), *Handbook on Business Process Management 1*. International Handbooks on Information Systems. Berlin, Heidelberg: Springer-Verlag.

Hammer, M., & Champy, J. (1993). *Reengineering the Corporation: A Manifesto for Business Revolution*. New York: HarperCollins.

Harrington, H. J. (1991). *Business Process Improvement: The Breakthrough Strategy for Total Quality, Productivity and Competetiveness*. New York: McGraw-Hill.

Imai, M. (1986). *Kaizen: The Key to Japan's Competitive Success*. New York: McGraw-Hill.

Jeston, J., & Nelis, J. (2014). *Business Process Management*. London and New York: Routledge.

Jurczuk A. (2018). Identification and Tracking of Process Inconsistencies in Manufacturing Enterprises. In S. Sahin (Ed.), *8th International Conference on Engineering, Project, and Product Management (EPPM 2017)*. Lecture Notes in Mechanical Engineering (pp. 129–138). Cham: Springer.

Kosieradzka, A., Kąkol, U., & Krupa, A. (2011). The Development of Production Management Concepts. *Foundation of Management, 3*(2), 55–74.

Lee, R. G., & Dale, B. G. (1998). Business Process Management: A Review and Evaluation. *Business Process Management Journal, 4*(3), 214.

Linderman, K. K., Schroeder, R. G., & Sanders J. (2010, November). A Knowledge Framework Underlying Process Management. *Decision Sciences, 41*(4), 689–719.

Lunau, S. (Ed.). (2008). *Six Sigma + Lean Toolset*. Berlin, Heidelberg: Springer-Verlag.

Malinova, M., & Mendeling, J. (2018). Identifying do's and Don'ts Using the Integrated Business Process Management Framework. *Business Process Management Journal, 24*(4), 882–899.

McCormack, K. (2001). Business Process Orientation: Do You Have It? *Quality Progress, 34*(1), 51–58.

Nadarajah, D., Syed, A. & Kadir, S. L. (2016). Measuring Business Process Management Using Business Process Orientation and Process Improvement Initiatives. *Business Process Management Journal, 22*(6), 1069–1078.

Nadler, G. (1967). *Work System Design the Ideal's Concept*. Homewood: Irwin.

Netjes, M., Reijers, H., & van der Aalst, M. (2006). Supporting the BPM Life-Cycle with FileNet. In *Proceedings of the Eleventh CAiSE/IFIP8.1 International Workshop on Evaluation of Modeling Methods in Systems Analysis and Design* (pp. 497–508).

Oakland, J. S. (2014). *Total Quality Management and Operational Excellence*. New York: Routledge.

Pande, P. S., Neuman, R. P., & Cavanagh, R. R. (2001). *The Six Sigma Way: How GE, Motorola, and Other Top Companies Are Honing Their Performance*. New York: McGraw-Hill.

Porter, M. E. (1985). *Competitive Advantage: Creating and Sustaining Superior Performance*. New York: Free Press.

Rosemann, M., & vom Brocke, J. (2015). The Six Core Elements of Business Process Management. In J. vom Brocke & M. Rosemann (Eds.), *Handbook of Business Process Management 1* (pp. 107–122). Berlin, Heidelberg: Springer.

Rummler, G. A., & Brache, A. P. (1995). *Improving Performance*. San Francisco: Jossey-Bass Publishers.

Van der Aalst, W. (2013). *Business Process Management: A Comprehensive Survey*. In International Scholarly Research Notices, Volume 2013, Hindawi.

Womack, J. P., & Jones, D. T. (1996). *Lean Thinking: Banish Waste and Create Wealth in Your Corporation*. New York: Simon & Schuster.

Womack, J. P., Jones, D. T., & Ross, D. (1990). *The Machine That Changed the World*. New York: Macmillan.

Ziemba, E., & Obłąk, I. (2014). Informatyczne wsparcie procesów w administracji publicznej. *Roczniki Kolegium Analiz Ekonomicznych/Szkoła Główna Handlowa, 33,* 619–646.

Zur Muehlen, M., & Ho, D. T. (2006). Risk Management in the BPM Lifecycle. In C. Bussler & A. Haller (Eds.), *Business Process Management Workshops* (pp. 454–466). Berlin, Heidelberg: Springer.

2

The Process Lifecycle

Katarzyna Rostek

2.1 Overview of Lifecycle Examples—Katarzyna Rostek

The implementation of process management in an organization most often relies on the PDCA (Plan - Do - Check - Act) cycle, but the exact interpretations of the same can differ depending on the specificity of respective approaches.

2.1.1 Three Phases

The simplest approach is where three basic stages of the PDCA cycle are distinguished, i.e. Plan-Do-Check, with Action treated as the output of Check in the previous iteration and the input to Plan in the subsequent iteration. Presented below are two examples of such an approach. The three-phase process lifecycle model is shown in Fig. 2.1. It consists of the following phases (Nurcan et al. 2012, pp. 231–236): (1) design, (2) deployment, (3) operation and evaluation. The model is very close to the

A. Kosieradzka and K. Rostek, *Process Management and Organizational Process Maturity*,
https://doi.org/10.1007/978-3-030-66800-6_2

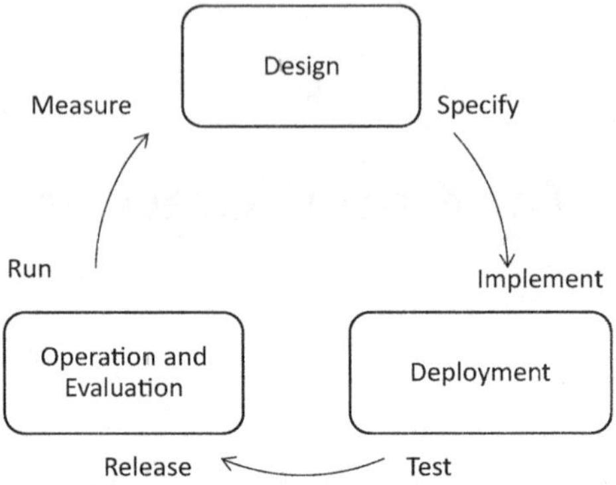

Fig. 2.1 The three-phase process lifecycle (*Source* Nurcan et al. 2012, p. 232)

classic PDCA, but it is distinguished by including the IT context in its operation.

The design phase begins with the definition of requirements and ends with the implementation of the process. The process is tested against the customer's requirements in the deployment phase. After successful testing, the process is released to be included in operations. Next, the process is measured and analyzed to identify problems and potential improvements in the operation and evaluation phase. These problems are the input to the next iteration of the lifecycle and the phase of process redesign.

Notably, IT technology is actively utilized in each of the above phases to facilitate effective identification, analysis, and eventual redesign of new process variants. The multitude of approaches and diversity of proposed tools additionally strengthen the synergy effect between technology and process maturity in this model. This approach should be effective for well-structured processes with high operations stability and high susceptibility to automation and digitization. But it may prove insufficient in the case of processes characterized by high variability, both procedural and functional, for which the full evaluation and improvement phase is needed to decide on the areas and scopes that should be changed.

Figure 2.2 shows the redesigned BPM lifecycle (Van der Aalst 2013, pp. 1–12). In the (re)design phase, a process model is identified and described, after which the run and adjust phase starts. In this phase, the

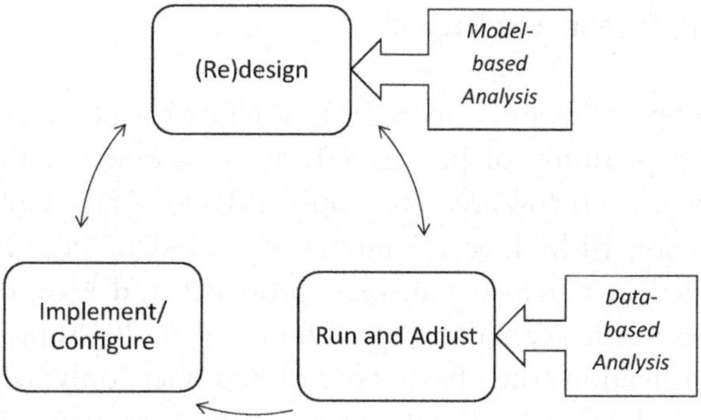

Fig. 2.2 The redesign BPM lifecycle model (*Source* Van der Aalst 2013, p. 5)

process model is enacted and adjusted in accordance with the needs and possibilities. In that phase the process is not redesigned, but only predefined control measures are used to adapt or reconfigure the same. At the same time, it is still possible to return to the redesign phase and introduce deep changes to the process. This model is transformed into an executive stage in the implementation/configuration phase.

The redesign BPM lifecycle entails two types of analyses: model-based and data-based. Current data are collected while the system is running. These data can be used to analyze processes that are already running or being adjusted, for example to discover bottlenecks, wastages, and deviations. This is the input for the redesign phase. In this phase many analytical models can be used in the analysis. For example, a very popular approach involves "what-if" simulations or an analyses verifying the correctness and effectiveness of a new project before its implementation and deployment.

Indication of the correlation between process phases and types of supportive analyses is a very important distinguishing feature of this model. Therefore, it will be particularly useful in situations where the processes are subject to frequent changes, continuously monitored, measured and evaluated, and the organization pays great attention to their optimization and improvement.

2.1.2 First Phase Excluded

The next group of lifecycles proposes to exclude the phase of initial identification and planning of processes from subsequent iterations of the lifecycle. Presented below are two examples demonstrating this approach.

The six-phase BPM lifecycle model presented in Fig. 2.3 illustrates phases entailed in managing business processes and ways of improving said processes in the six major steps (Dumas et al. 2018, pp. 22–26): (1) process identification (the phase is performed once, only in the first iteration of the cycle), (2) process discovery, (3) process analysis, (4) process redesign, (5) process implementation, and (6) process monitoring and controlling.

Process Identification is concerned with setting up a high-level description of the organization's main processes and an assessment of their current state. The primary result of this phase is a map of process architecture which identifies the organization's main processes, describes the relationships between the same, and defines criteria for their respective

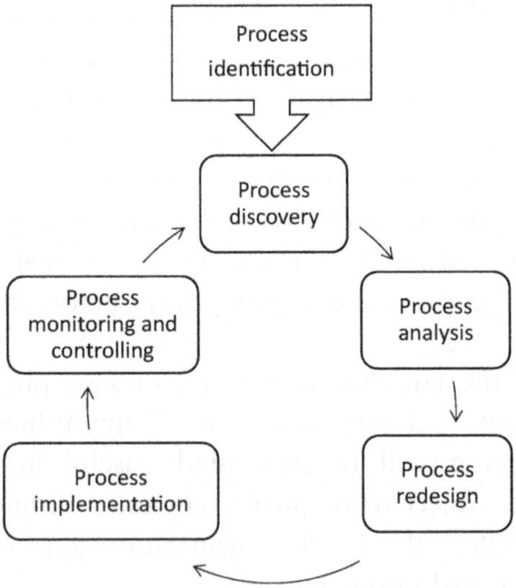

Fig. 2.3 The six-phase BPM lifecycle model (*Source* Dumas et al. 2018, p. 23)

prioritization. The identified structure will be subject to further elaboration, activation and monitoring in subsequent stages of the repeatable and multi-iterative cycle.

After process discovery, the focus of the lifecycle shifts to a single, specific process. The process discovery phase produces a detailed description of the process in its current state. This description is referred to as an "as-is" process model. Process analysis applies analytical tools and techniques to determine weaknesses in the "as-is" state and identify the impact of each weakness.

Process redesign addresses the most important weaknesses of the process and produces a reworked model—the "to-be" process model. This model is subsequently used as the basis for process implementation. Process implementation typically entails executable process modeling, implementation of a supporting information system, and identification of process measures.

In the phase of process monitoring and controlling, execution data are continually collected and compared to the applicable performance and conformance objectives. Deviations from these objectives as well as changes in the business environment or the company's goals trigger a new iteration of the BPM lifecycle.

The six-phase BPM model assumes that the basic structure of organizational processes (called a processes map) is rather constant, so it is not necessary to include the identification phase in subsequent iterations.. Indeed, this is true in many cases, hence this model can be used in a wide range of situations.

The service-oriented business process development methodology (Papazoglou and Van Den Heuvel 2007) is based on a roadmap that comprises one preparatory phase to plan development, and five distinct phases that concentrate on business processes: (1) analysis and design, (2) construction and testing, (3) provisioning, (4) deployment, and (5) execution and monitoring (Fig. 2.4).

Progression through the phases tends to be incremental and iterative and should account for the potential need for modifications in situations where the actual scope cannot be completely defined a priori. This allows for an approach that is one of continuous identification, discovery, design and implementation in each iteration. This approach considers multiple

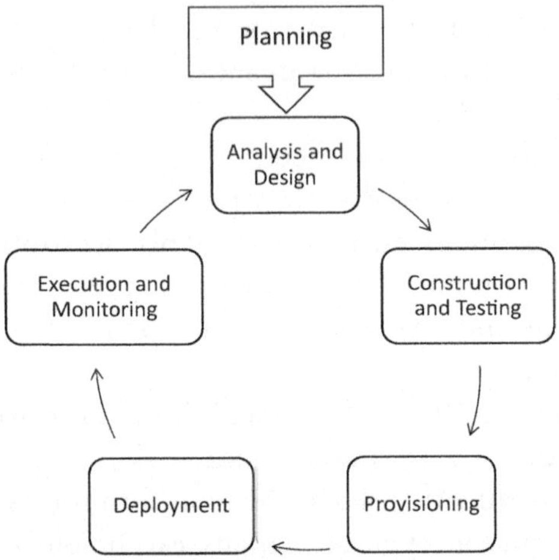

Fig. 2.4 The business process development lifecycle methodology (*Source* Papazoglou and Van Den Heuvel 2007, p. 80)

realization scenarios for business processes that take into account both the technical and business perspective. This is particularly noticeable in the case of Provisioning and Deployment phases which offer opportunities for optimization of not only the current process, but also associated activities and ancillary processes.

The planning phase determines the scope of activities in all subsequent phases. The decision is preceded by an analysis of the business case and probability of success for all possible combinations of development approaches and the associated implementation strategies. In the support technics area, this phase utilizes gap and scenario analyses. The analysis and design phase specifies processes in terms of:

- analyzing customer needs with regard to goods and services,
- identifying processes correlated with the created products and services,
- determining objectives and establishing the business process structure,
- describing process-related responsibilities (roles),
- identifying non-functional process concerns connected with efficiency or productivity.

The construction and testing phase requires the implementation of process models and then testing them in terms of functional and non-functional requirements identified in the previous phase. This action is for ensure compliance of the executive process version with the assumed results. Tasks executed during the provisioning phase involve making choices for governance, certification, enrolment, auditing, metering, billing, and managing operations that control the behavior of a process products and services during its exploitation and use. These activities support organization and its customers in future cooperation and communication.

The deployment phase includes every activities connected with implementation and operation of executive environment—organizational, business and technical. The execution and monitoring phase includes processes operation and control. This is the last phase of the cycle, followed by another cycle iteration, implementing a new range of changes and improvements in processes.

The service-oriented business process development methodology is focused on the implementation of processes related to the services provision, in particular web services. Therefore, it can be successfully used in organizations for which the Internet is the primary environment for operating and contacting customers. It will also work well in software development and other kind of operations very strongly focused on the customer and creation value for him.

2.1.3 Multiphase

Another category includes lifecycles whose implementation was elaborated to include numerous stages justified by the particular context of application.

The first of the same is dedicated to the environment of public administration (Batko 2009, pp. 53–64). It comprises nine stages (Fig. 2.5) the improvement of which relies on utilizing knowledge regarding problems encountered in the preceding iteration of the cycle.

The cycle commences with the stage of Process analysis and identification. In the course of various studies and analyses, products and services

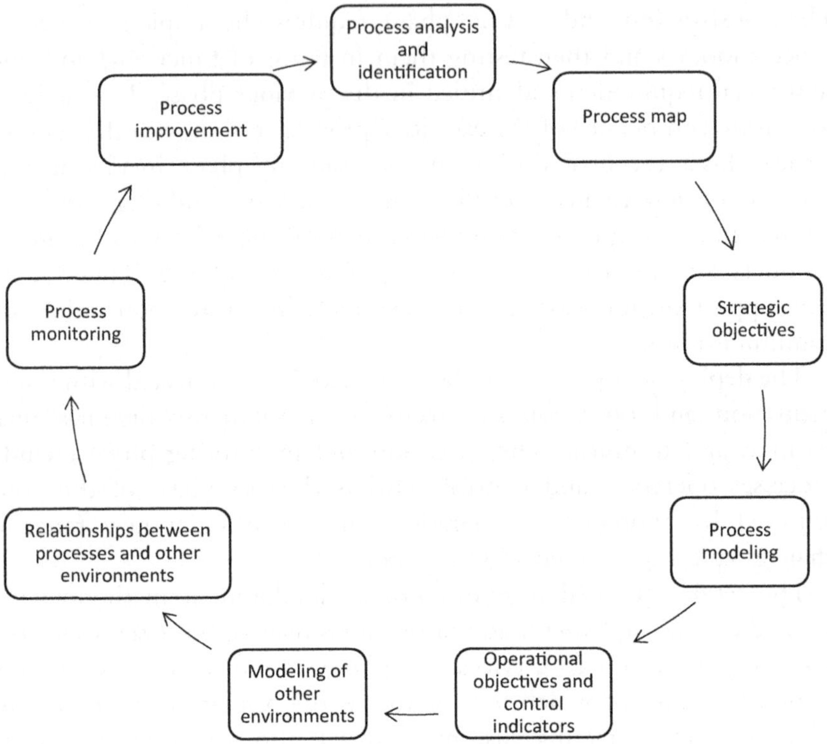

Fig. 2.5 Administrative process lifecycle (*Source* Batko 2009, p. 61)

are identified that will serve as the basis for respective processes, hence the processes pertinent to the given organization are identified. Following up on the same, a process map is developed (Process map phase) as a graphical representation of all organizational processes at a high level of generality. In the next stage, compatibility between processes and the organization's strategic objectives is verified (Strategic objectives phase). Processes should effectively facilitate the performance of corresponding strategic objectives.

At the stage of process modeling, processes are further elaborated (Process modeling phase). Each model includes descriptions of respective actions, decision-making pathways, activity parallelism, etc., which facilitates the correct and optimized execution of respective tasks, as well as allows better communication and more effective process risk management. Monitoring of process performance is conditioned by the operating objectives defined during the preceding phase and the adopted control indicators.

One of the major benefits of implementing process management in organizations stems from the integration of various intraorganizational environments, e.g. associated with labor, material and human resources, IT, or documentation. In the process lifecycle, these will be used as foundations in the establishment of references for the benefit of process models. This starts at the sage of Modeling of other environments and is further elaborated in the phase of Relationships between processes and other environments. References take into account scopes of responsibility, roles, and characteristics that effectively facilitate coordination and integration of the given process with the environments affecting it. The stages of process monitoring and process improvement complete the lifecycle by creating conditions for optimization and perfection. The stages are heavily reliant on the control indicators and results of operating processes' execution.

A characteristic trait of this cycle is its association with the functional performance of the entire organization and its elements that at first glance seem unaffected by the respective processes, e.g. the given external environments (legal, social, demographic, economic). Consequently, the cycle is particularly applicable to organizations whose environment and stakeholders are vital to operational performance, as well as ones wherein the influence of the same is strong enough not to be omissible at the stage of process design and monitoring.

The next example is the process lifecycle in the LEADing Practice model integrating the process lifecycle with the areas of management, tool application, and responsibility for the process in a single common conception (Fig. 2.6). For the sake of clarity, only the process context of the model will be described hereinbelow (Scheruhn et al. 2015, pp. 511–550).

The analysis phase is the first step in providing a complete discovery of existing (as-is) processes, closely followed by the capture, decomposition, and documentation of all relevant related information objects, properties, and relationships. The processes, together with the related

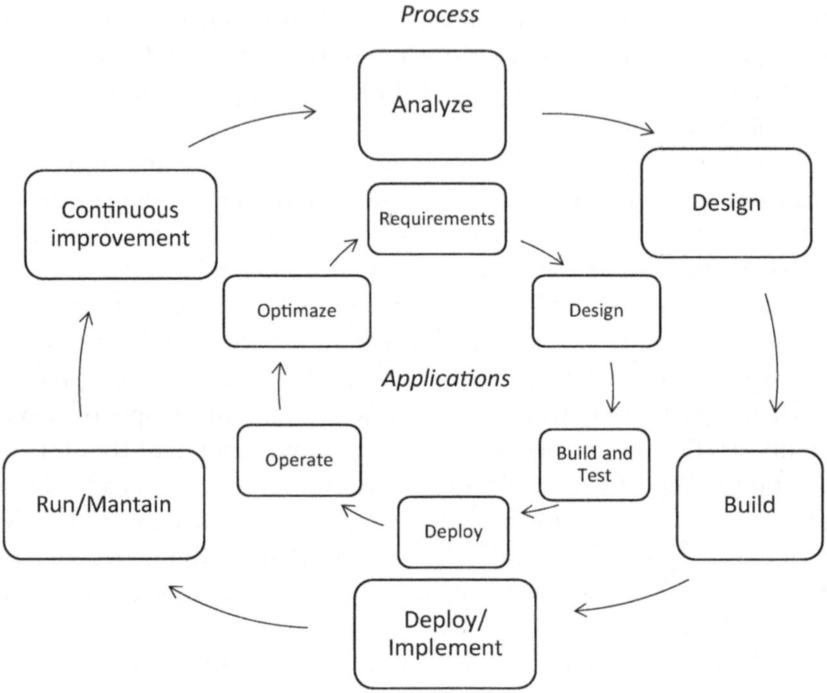

Fig. 2.6 The process lifecycle in the LEADing Practice model (*Source* Scheruhn et al. 2015, p. 513)

dynamic and static business structures, should ultimately support and execute the strategic objectives and critical success factors of the organization. These strategic aspects are a part of the business expectations and business requirements that have to be considered in the analysis and organizational design of the associated strategic information objects and their relationships to the processes.

In the design phase, the new process model and structure are designed (to-be status). Depending on the scope of the project, the design work can include anything from altering the complete process construction, through adding or deleting processes, to only minor changes in the basic process structure. A similar range of the changes can occur in connection with the information objects and the related dynamic and static structures contained within each of the processes.

The Build phase is concerned with applying the "to-be" models defined during the Design phase, including all related dynamic and static structures, needed to create the operating system (manual or automated). In the manual situation, the Build phase is focused on work design, training, and preparation of documentation. Where an automated environment is established, the activities may also include programming, configuration, or other software tasks that perform the process or enable its performance. Obviously, in many cases both types of work (manual and automated) will be required and must be coordinated to complete or achieve the results that expected of the new process iteration.

In the Deployment phase, processes based on the "to-be" models are implemented and executed. The process models and the information they convey can serve as the basis for testing and can be used to offer a high level of support during the implementation.

The Maintain phase is concerned with processes operating performance and its enablers in the execution environment. During this phase, efforts must be made to ensure that the process operation remains consistent with the design objectives. This is also the main task of the final, Process monitoring phase, which ultimately provides the input for the analysis phase in the next iteration of the cycle.

Once the new processes are operational, ongoing work is necessary to verify whether the intended goals have been met through a continual effort to learn from and improve on the design of the process to achieve its design goals. These efforts can seek to implement evolutionary changes or may involve innovative changes to the design. Continuous improvement is a key aspect of BPM whereby feedback from the process and the customer is evaluated against design goals. This cycle is especially dedicated to those organizations in which the process environment is largely automated and supported by dedicated IT tools and applications. In such a situation, it is extremely important to coordinate the software life cycle (or even the entire corporate architecture) and the process, which enables the presented approach.

2.1.4 Pointed Dependencies

The last characterized group includes models whose function, although based on the lifecycle, is definitely more complex and takes into account the interdependence or/and internal complexity of particular phases.

The example of such a model (Weske 2012, pp. 11–17) is presented in Fig. 2.7. It consists of interrelated phases organized in a cyclical structure, showing their logical dependencies. These dependencies do not imply a strict temporal ordering in which the phases need to be executed. Many design and development activities are conducted during each of these phases, with characteristic incremental and evolutionary approaches involving concurrent activities in multiple phases.

Processes and their organizational and technical environments are studied in the design and analysis phase. Based on this research, processes are identified, reviewed, validated, and represented by process models. Process modeling is the core technical subphase during process design. Based on the findings of the process improvement activities, informal

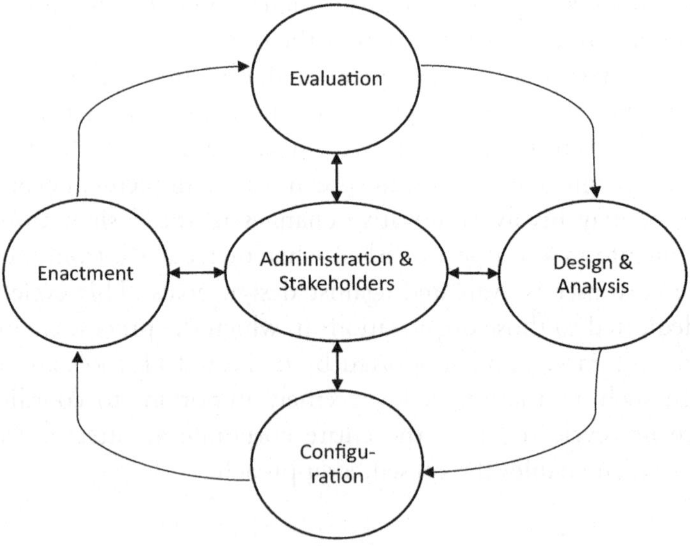

Fig. 2.7 The business process lifecycle with dependent phases (*Source* Weske 2012, p. 12)

process descriptions are formalized using a particular process modeling notation.

Once the process model has been designed and verified, the process needs to be implemented. There are different ways to approach this task. Implementation can be accomplished through a set of policies and procedures that the employees need to comply with, notably without any support from a dedicated process management system. If a dedicated software system is used to realize the processes, an implementation platform is chosen during the configuration phase. In this case, the process model is enhanced with technical information that facilitates the process enactment by the process management system.

The enactment phase encompasses the real run-time of the process. Process instances are initiated to fulfill organizational goals. The process management system actively controls the execution of process instances in accordance with the definition of the process model. Process enactment needs to cater to a correct process configuration, guaranteeing that the process activities are performed according to the execution constraints specified in the process model. The evaluation phase uses available information to evaluate and improve process models and their implementations. Techniques used in this phase are aimed at identifying the quality of process models and their adequacy to the execution environment.

The administration and stakeholders' stage is located halfway through the cycle. There are numerous artefacts emerging at different levels of abstraction that need to be organized and managed. Example elements that should be taken into account can include structured storage and efficient retrieval of artefacts regarding process models and information on process instances as well as the organizational and technical execution environments. Additionally, the process domain is characterized by several types of stakeholders with different knowledge, skills and experience, classified under specific process roles or functions and ensuring high efficiency of execution. Such an organization is primarily intended to properly coordinate the remaining phases of the model.

Placing administrative functions and process stakeholders in the center of the model means that the cycle will fit well with processes typical of public administration units, where the impact of administrative functions should be well correlated with customer needs.

The BPM CoE process lifecycle model is controlled, directed, and governed by the BPM Center of Excellence (von Rosing et al. 2015, pp. 219–243) across all areas of the business. It consists of six phases (Fig. 2.8): analyze, design, build, deploy/implement, run/maintain, and continuously improve.

Although the model's phases are the same as those shown in Fig. 2.6, what is important is that it applies and is implemented in relation to the entire portfolio of processes whose execution is to achieve specific benefits and aims. Therefore, its implementation requires the involvement of a number of people with often unique competencies who are able to design and implement a cross-sectional change that concerns many processes and affects many organizational activities simultaneously. Such persons are employed by a specialized unit known as the BPM Center of Excellence. The purpose of creating and maintaining such a unit is to ensure continuous improvement of processes, but with due consideration for the needs, expectations and goals of the entire organization, its environment and, above all, its stakeholders. It should also be noted that the

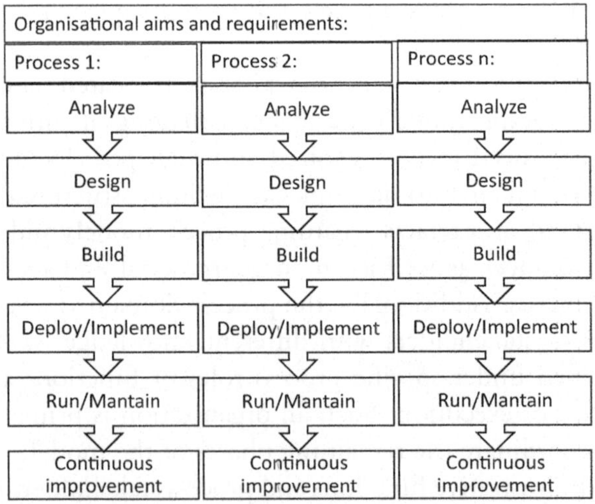

Fig. 2.8 The BPM CoE process lifecycle (*Source* Own elaboration)

BPM CoE is not focused on a specific process, but rather on the goals of the organization. Hence, it is not iterative (improving each process individually and in isolation from others) but project-based (projects are deployed to achieve specific goals or change specific scopes in relation to all the processes involved).

The proposed model can certainly be applicable in organizations where the impact of the organizational structure's complexity, functionality (processes), the environment, and stakeholders is so strong and multidimensional that it can constitute a serious organizational and project-related challenge. International organizations with a diversified product or service offer tend to fall under this category.

2.2 Results Analysis—Adaptation of the PDCA Cycle—Katarzyna Rostek

The analysis considered four groups of process lifecycle models:

- simplified cycles, where the stage of evolution and improvement is not treated as a separate element but rather included as a part of the implementation or design stages,
- cycles where the stage of process planning and identification is treated separately and performed only once, i.e. not repeated in subsequent iterations of the cycle,
- detailed cycles, where a greater number of stages is intended to more strongly correlate process management with other areas of management (e.g. IT, HR),
- cycles with elements of co-dependence, where the respective stages not only and not always add up to a chronological cycle, but also create a network of interdependences and subordinations that may alter the original chronology with a view to achieving higher-level and more complex organizational goals.

Upon analyzing the presented examples and having preliminarily classified the same, one arrives at the following conclusions:

- at the input, the process lifecycle is modeled after the classical PDCA cycle, and subsequently adapted to the individual needs and nature of the given application,
- the process lifecycle should be correlated (account for the co-dependence) with the following aspects: organization (corporate architecture), environment (market, economy, law), technology (IT infrastructure, systems and support tools), and people (internal and external stakeholders),
- the lifecycle ought to be correlated with the lifecycles of the supporting IT tools and projects whose goal is to implement new or modify existing processes,
- there is no such thing as a standard and unchanging process lifecycle as it is defined by the comprehensive organizational approach to process management which should evolve in time as new experience is gathered,
- on the other hand, there are a number of factors indicating that the specific types of lifecycles ought to be adapted and dedicated to the reality of the given organization. The same are the most strongly associated with the type of activity being conducted, size and organizational structure, as well as the environment affecting the organization.

The above leads to the conclusion that a process lifecycle ought to be an element of an integrated organizational management model (enterprise integration), conditioned by the lifecycle of the enterprise's architecture. Should one consider processes from the perspective of an integrated model of organization-wide management, it becomes considerably easier to understand the complexity of various correlations and dependencies that result from the identification, isolation, and implementation of each comprehensive process life cycle. BPTrends Associates quite aptly illustrated the same in the following pyramid graph (Fig. 2.9).

Figure 2.9 shows that enterprise architecture is dedicated to achieving enterprise integration by capturing and describing processes, strategies, organizational structure, resources, goals and constraints of the enterprise (Bernus et al. 2012, p. 10). This approach to the project lifecycle yields a

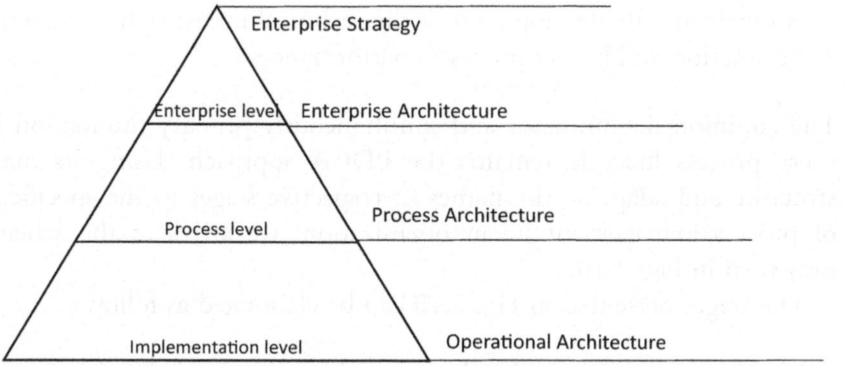

Fig. 2.9 Enterprise integration model (*Source* http://www.bptrendsassociates.com/about-us/bpm-methodology/ <read 2020-07-03>)

number of benefits, for instance in terms of (Woźniak and Fester 2015, pp. 209–210)[1]:

- understanding of the organization's specificity (possibilities, limitations, and requirements) in terms of process execution, notably in close relation to its goals, functional principles, policies, and availability of resources,
- ensuring the adequate correlation between the specific strategic goals and the solutions generated by the processes (products or services),
- encouraging multi-aspectual improvement of processes within the lifecycle, e.g. in terms of strategic planning or increased compatibility between long-term business and technological planning,
- indicating the need to deploy the processes of guiding innovation and improving value chains, including in terms of rationalizing IT project management,
- facilitating the creation, configuration, and improvement of communication channels networks with the stakeholders (both internal and external) as an element of streamlining processual flows of data and information,

[1]Woźniak, J., & Fester, A. (2015). Architektura korporacyjna–źródło korzyści dla przedsiębiorstwa. *Edukacja Ekonomistów i Menedżerów, 38*(4), 201–227.

– rationalizing the development of the technical infrastructure enabling automation and higher processes performance.

The common denominator and simultaneously primary foundation in every process lifecycle remains the PDCA approach. Using its main structure and adapting the names of respective stages to the specificity of process management in an organization, we arrive at the scheme presented in Fig. 2.10.

The stages presented in Fig. 2.10 can be elaborated as follows:

– Modeling—activities related to the identification, definition, and design of processes, as well as modification of already existing processes (redesign),
– implementation—activities related to organization of the operating environment, provision of necessary means and resources, deployment and execution of processes,
– verification—activities related to continuous monitoring, control, and documentation of process results (including elements and areas that require improvement),
– perfection—activities related to solving problems, improving and developing processes, which provide the basis for determining the

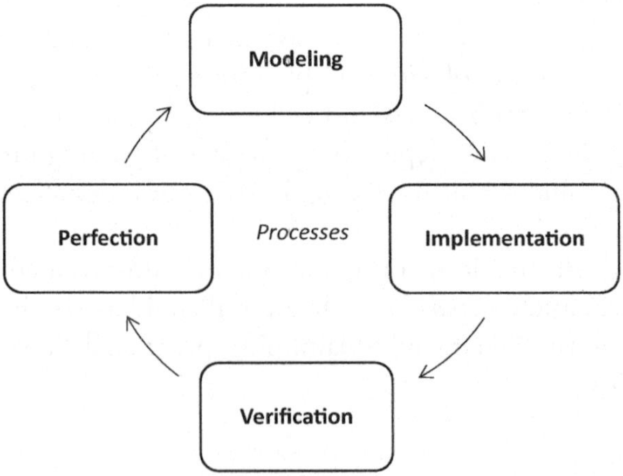

Fig. 2.10 The basis business process lifecycle (*Source* Own elaboration)

scope of necessary changes as well as planning their implementation in the subsequent iterations of the process lifecycle.

The proposed structure will be used in the presentation of detailed elaborations on the respective stages of the cycle considered in the context of economic organizations (marketing products or services) and non-economic organizations (public administration). The goal is to demonstrate that there are numerous elements and factors that need to be taken into account in effective process management if it is to contribute not only to the efficiency of present operations but also their future development. Therefore, apart from indicating the discrepancies observed in the context of executing each respective stage, certain recommendations will also be provided with a view to further improving the effectiveness of process management in various types of organizations.

References

Batko, R. (2009). Urząd gminy jako organizacja procesowa. *Zarządzanie publiczne. Zeszyty Naukowe ISP UJ, 4*(8), 53–64.

Bernus, P., Nemes, L., & Schmidt, G. (2012). *Handbook on Enterprise Architecture*. Berlin and Heidelberg: Springer Science & Business Media.

Dumas, M., La Rosa, M., Mendling, J., & Reijers, H. A. (2018). *Fundamentals of Business Process Management* (2nd ed.). Berlin: Springer.

Nurcan, S., Schmidt, R., Bider, I., & Regev, G. (2012). Special Issue on Business Process Lifecycle: Design, Deployment, Operation and Evaluation. *Journal of Software: Evolution and Process, 24*(3), 231–236.

Papazoglou, M. P., & Van Den Heuvel, W. J. (2007). Business Process Development Life Cycle Methodology. *Communications of the ACM, 50*(10), 79–85.

Scheruhn, H. J., von Rosing, M., & Fallon, R. L. (2015). Information Modeling and Process Modeling. In M. Von Rosing, H. Von Scheel, & A. W. Scheer (Eds.), *The Complete Business Process Handbook: Body of Knowledge from Process Modeling to BPM, Volume 1* (pp. 511–550). Waltham, MA: Morgan Kaufmann.

Von Rosing, M., Hove, M., von Scheel, H., & Morrison, R. (2015). BPM Center of Excellence. In M. von Rosing, H. von Scheel, & A. W. Scheer (Eds.), *The Complete Business Process Handbook: Body of Knowledge from Process Modeling to BPM* (Vol. 1, pp. 219–243). Waltham, MA: Morgan Kaufmann.

Weske, M. (2012). *Business Process Management: Concepts, Languages, Architectures*. Berlin: Springer.

Woźniak, J., & Fester, A. (2015). Architektura korporacyjna–źródło korzyści dla przedsiębiorstwa. *Edukacja Ekonomistów i Menedżerów, 38*(4), 201–227.

Van der Aalst, W. (2013). *Business Process Management: A Comprehensive Survey*. In International Scholarly Research Notices, Volume 2013, Hindawi. https://doi.org/10.1155/2013/507984.

3

Process Modeling

Anna Kosieradzka and Katarzyna Rostek

3.1 Process Modeling in Production Companies—Anna Kosieradzka

3.1.1 The Context of Production Companies

A company is a legally, organizationally, and economically identified entity conducting business activity. The essential function of a company is to manufacture products and/or render services. It is accepted that the short-term objective of a company is to generate profit by satisfying consumer needs, whereas mid and long-term goals pertain to the increase of its value to benefit its owners.

In order to facilitate the realization of those objectives, decisions have to be made in the following functional areas of the company (Table 3.1).

In light of the systems theory (see Sect. 1.2), a company can be perceived as a production system consistent with the model presented in Fig. 3.1. The model has been employed in various publications by authors subscribing to the systemic approach.

A. Kosieradzka and K. Rostek, *Process Management and Organizational Process Maturity*,
https://doi.org/10.1007/978-3-030-66800-6_3

Table 3.1 Areas and primary decisions made in the context of managing production

No.	Ten decision areas	Issues
1	Design of goods and services	What goods or services should we offer? How should we design these products?
2	Managing quality	How do we define quality? Who is responsible for quality?
3	Process and capacity design	What process and what capacity will these products require? What equipment and technology is necessary for these processes?
4	Location strategy	Where should we put the facility? On what criteria should we base the location decision?
5	Layout strategy	How should we arrange the facility? How large must the facility be to meet our plan?
6	Human resources and job design	How do we provide a reasonable work environment? How much can we expect our employees to produce?
7	Supply-chain management	Should we make or buy this component? Who should be our suppliers and how can we integrate them into our strategy?
8	Inventory, material requirements planning and JIT (just-in-time)	How much inventory of each item should we have? When do we reorder?
9	Intermediate and short-term scheduling	Are we better off keeping people on the payroll during slowdowns? Which job do we perform next?
10	Maintenance	How do we build reliability into our processes? Who is responsible for maintenance?

Source Heizer and Render (2011, p. 39)

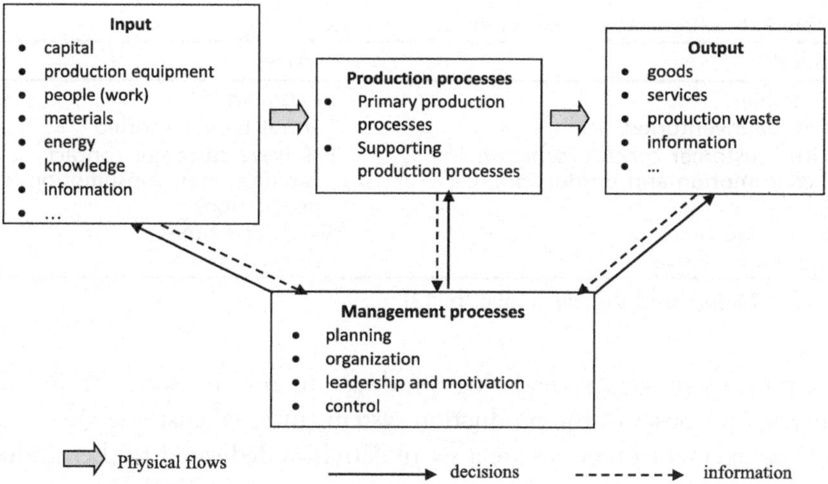

Fig. 3.1 Production system model (*Source* Own elaboration)

The outputs of a production system are products, i.e. goods or services produced by the system and sold to customers. The outputs of production systems also include the byproducts of production processes, such as usable wastes that may be resold for utilization (e.g. scrap metal, steam, etc.) as well as harmful wastes that generate additional costs related to their neutralization or payment of applicable environmental fees.

Products can take the form of goods or services. The key differences between the same, significant from the perspective of identifying processes emerging in production management, are presented in Table 3.2.

The inputs of production systems include all tangible and intangible production factors necessary for the manufacture of the outputs, i.e. the system's products. They include: financial capital, production equipment, people (work), materials, energy, knowledge, information, etc.

Production processes should be understood as the body of activities aimed at converting the system's inputs into products, i.e. outputs. Two main types of production processes can be distinguished: primary processes, constituting a sequence of activities through which products intended for sale (to external customers) are manufactured, and

Table 3.2 Characteristics of goods and services

Goods	Services
Tangible	Intangible
Can be inventoried	Cannot be inventoried
Little customer contact (separate consumption and production)	Extensive customer contact (simultaneous consumption and production)
Long lead times	Short lead times
Capital intensive	Labor intensive

Source Melnik and Denzler (1996, p. 13)

supporting processes rendering products (goods or services) for the internal purposes of the production system (internal customers).

Management processes are a set of activities dedicated to the production system's resources (human, financial, material, and information) and performed with a view to manufacturing specific products (goods or services) to effectively and efficiently satisfy the diverse needs of customers.

Relative to the way of satisfying the needs of external customers, two distinct types of goods production models can be distinguished: Make to Stock (MTS) and Make to Order (MTO), which differ in terms of process scope and position in the value creation chain. The model applicable to the production of services needs to be analyzed separately given the differences between goods and services indicated in Table 4.2.

Production to stock (MTS—Make to Stock) refers to a situation where orders are completed from stock reserves and production is expected to resupply said stock. Products are manufactured before receiving orders from customers, based on demand forecasts. Production orders are initiated when the stock of given finished product reaches the reorder point. This models is typical of stable mass production or medium- and large-batch production of unified products, usually from the groups of so-called fast moving consumer goods (FMCG), e.g. food, pharmaceuticals, cosmetics, domestic detergents, minor household products, as well as furniture, electronics and household appliances. More advanced versions of the model are sometimes referred to as Make to Availability (MTA) to emphasize that the primary goal of the system is to ensure high availability of the product for customers.

Production to order (MTO) is performed exclusively as ordered and required by the customer, rather than based on sales forecasts. Characteristically, in this system the product can be manufactured only once an order has been placed by a customer. Two distinct variants can be identified in this context:

- a product is designed and manufactured in accordance with the customer's specifications (e.g. a building, a ship, etc.); in which case the production cycle includes phases of design, manufacture, and project management;
- a product is selected by the customer from the producer's standard offer and adapted to the given customer's needs and requirements (e.g. ERP IT systems configured to the customer's specification or a machining tool adapted to the conditions at the customer's facility).

This model is characteristic of non-stable, either unit-based or small-batch manufacturing. Production is initiated upon receiving an orders from a customer. In order to determine a realistic lead time, one must be familiar with the product's production cycle. The final product typically comprises components that can be manufactured in advance with a view to facilitating future orders, as well as elements (sub-assemblies) that are unique and dedicated to the specific (accepted) order. A particular variant of this model is known as Assembly to Order (ATO), wherein the planning of pre-assembly stages relies on forecasts, while planning of the finished products' assembly is done based on customer orders. The ATO model tends to be applicable to situations where a large selection of final products is available, and products are assembled from pre-manufactured, standardized components.

Notably, it is fairly common for the two models described above to occur in parallel within a single enterprise.

Given the differences between goods and services presented in Table 3.2, processes involved in the context of services will be described separately and limited—as already mentioned above—only to the range of production services. Production services are defined as all services rendered by economic entities to other economic entities that contribute to the process of production while not directly producing any new tangible goods.[1] The same include in particular:

- installation and assembly works;
- gas and power distribution;
- transport services rendered to production facilities (transport of materials and semi-finished products, product distribution);
- repair, renovation, and maintenance works rendered to third parties (excluding warranty repairs performed independently by the manufacturer);
- services rendered on commission from product manufacturers, entailing the performance of a special technical procedure on an object supplied by the customer, e.g.: forming, coating, assembly of supplied components.

The specificity of production services stems from the very character of the same. They typically do not involve or only marginally involve storage processes—these may be necessary for instance in the case of a special technological procedure performed on supplied material that needs to be stored.

3.1.2 Process Identification

As already mentioned in Chapter 1, the basis for process management was formulated in early twentieth century within the classical school and stemmed from the character of production processes as such. Before a production company can manufacture a product, it must first complete the subsequent steps of technical preparation presented in Fig. 3.2.

At the stage of technical production preparation, the technological process is designed to serve as the starting point for the identification of the primary and supporting production processes. Apart from technological processes, production processes also include control, maintenance, transport, and storage processes. A technological operation occurs when

[1]Cabinet of Ministers Regulation of October 29, 2008 on the Polish Classification of Goods and Services (PKWiU), Dz. U. [Journal of Laws] No. 207, item 1293, as amended.

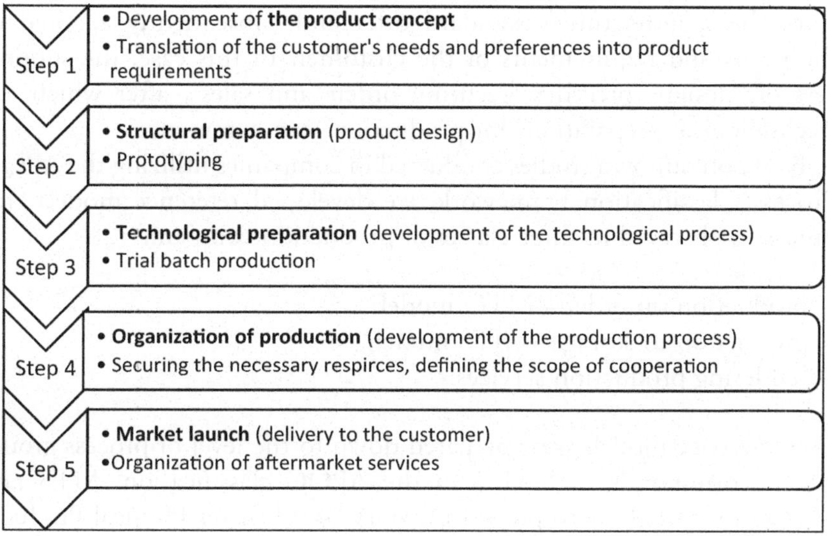

Fig. 3.2 Stages of technical production preparations (*Source* Own elaboration)

a change of the dimensions, shape, physical or chemical properties of the input material or semi-finished product takes place. Assembly (or disassembly) constitutes a particular type of a technological operation that entails modifying a part's position relative to other parts.

Control involves verifying the correctness of modifications to objects introduced due to deliberate processing or natural processes.

Transport entails the movement of the processed object from one location to another.

Maintenance involves securing the processed material against loss of the properties gained in the technological process.

Storage refers to the dormant condition of the processed object placed within a designated warehouse, immediate storage place, or directly on technological, transport, or control equipment. Wait times are also considered as elements of the storage operation.

In the MTO model, the process of product design and technological preparation for production may at times be divided into two separate substages of "design" and "technological preparation". This is applicable in situations where the product ordered by the customer is selected from the manufacturer's standard offer and subsequently adapted to the needs and requirements of the customer. In this case, the subprocess of "design" precedes "securing orders and sales", after which the "technological preparations for production" can commence.

Based on our own studies conducted in companies utilizing the APQC Process Classification Framework, we developed reference models[2] for processes observed in three classes of production companies:

- production on order—MTO model,
- production to storage—MTS model,
- rendering production services.

The reference models were prepared down to the level of process group, which corresponds to level 2 in the APQC classification. MTS and MTO models belong to process category "4.0 Deliver Physical Product" while the model for services can be classified as "5.0 Deliver Services". Table 3.3 characterizes the respective process groups and processes under the "Deliver Physical Product" category. Figure 3.3 presents the structure of processes observed in the MTO model, while Fig. 3.4—the corresponding structure for the MTS model. The main differences between the two pertain primarily to the position of sales processes in the value creation chain. In the MTO model, the process of "securing orders and sales" is located at the beginning of the value chain, whereas in the MTS model—at the end, as in the latter customer orders are completed from product stocks manufactured in advance based on demand forecast.

The provision of production services entails a somewhat narrower scope of processes and process groups compared to manufacturing, which—as already mentioned—stems from the very nature of such activities. In the context of services, the planning process focuses on determining requirements in terms of production resources (employees,

[2] A reference model is an abstract framework for understanding significant relationships among the entities of some environment, and for the development of consistent standards or specifications supporting that environment (according to Organization for the Advancement of Structured Information Standards (OASIS), https://www.oasis-open.org/committees/soa-rm/faq.php, retrieved 3.08.2020.

Table 3.3 Characteristics of processes/process groups observed in product manufacturing

Process/process group no. (as per Fig. 3.3 and Fig. 3.4)	Name of process/process group	Process specifics
1	Securing orders and sales	The process of securing orders, particularly important in the case of products manufactured "to order" that need to be adapted to the specific requirements of the customer, during which the salesperson has to demonstrate knowledge in terms of the company's ability to manufacture the product
2	Design and technical preparation of production	The bulk of activities aimed at facilitating the conditions necessary to manufacture products; including development of the product's concept as well as structural, technological, and organizational preparation of the company for the given production process
3	Procurement of materials and utilities	Entails selection of suppliers, procurement of materials and components as well as utilities such as power, gas, and water, including the supply thereof to the company
4	Product manufacture	The process of processing materials and components to obtain the final product ordered by the customer—the process of creating customer value

(continued)

Table 3.3 (continued)

Process/process group no. (as per Fig. 3.3 and Fig. 3.4)	Name of process/process group	Process specifics
5	Product quality assurance	The process of controlling the quality of products and statistically directing production processes to guarantee the customer's satisfaction with the final product
6	Production logistics	Production logistics entails storage of materials until the moment of their supply to the production floor. In the production phase, logistic processes include internal transport, warehousing, and supply management
7	Distribution and aftermarket service	This group of processes includes calculation of the price and invoicing. Distribution logistics entails planning, storage of finished products, loading and delivery to end customers, and in some cases also installation of the product at the customer's premises
8	Repairs and maintenance	The processes of repair and maintenance of machines and other technical equipment to guarantee their uninterrupted availability (failure-free and efficient operation) in the course of performing production processes
9	Manufacture/purchase of tooling	Manufacture of tools and instruments for the purposes of own production processes, alternatively, purchase of the same from external providers

Process/process group no. (as per Fig. 3.3 and Fig. 3.4)	Name of process/process group	Process specifics
10	Supervision of monitoring and control equipment	The processes of repair and maintenance of monitoring and control equipment guaranteeing the desirable quality of products
11	Production strategy	Development of a strategy regarding introduction of new (improved) outputs into production, implementation of new technological processes, and development of the production capacity (investments, collaboration, hiring)
12	Management of the production capacity	Planning the development of production capacity, investments, replacement of depreciated equipment with new machines, etc. to facilitate the planned production tasks
13	Employee preparation	The processes involved in preparing employees for the performance of production tasks, in particular through training and adequate choice of incentives
14	Production planning and control	All activities aimed at determining the quantities and timeframes of production and ensuring that the plans are executed
15	Information systems management	Management of information and IT systems utilized in production, e.g. ERP, MES

Source Own elaboration

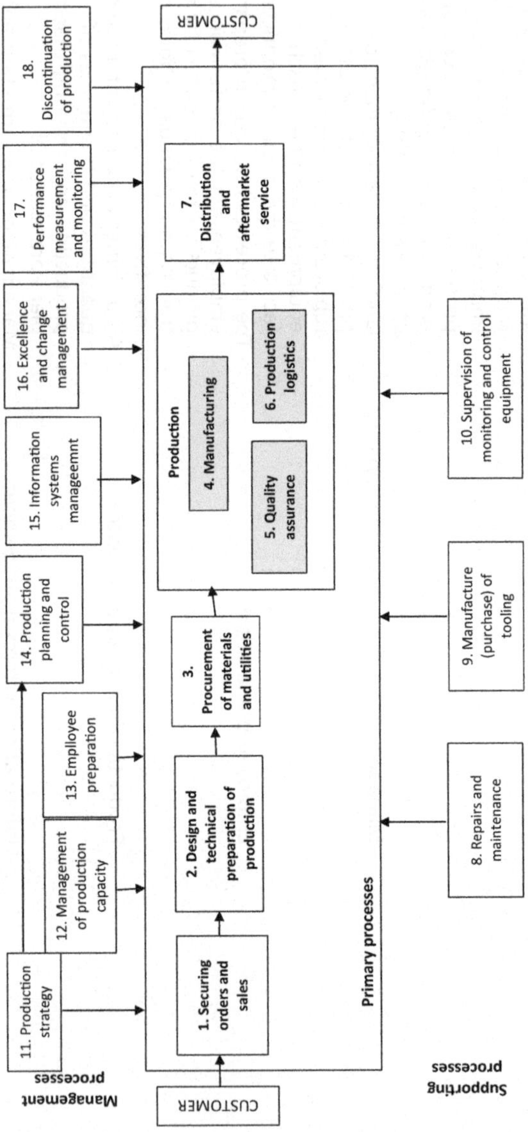

Fig. 3.3 Structure of processes in the MTO model (Make-To-Order) (*Source* Own elaboration)

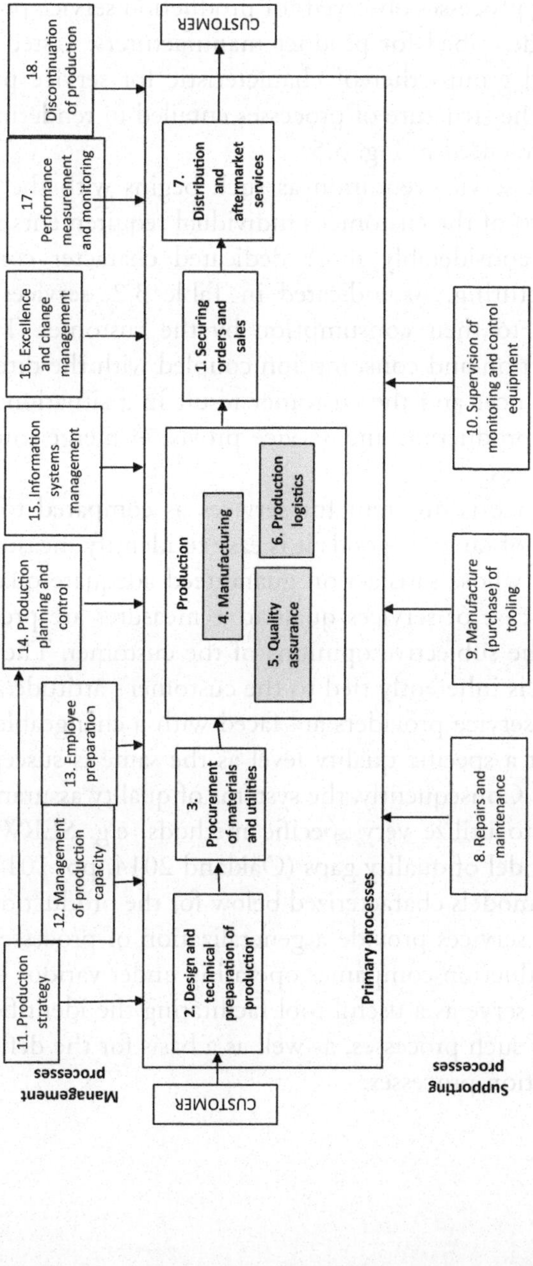

Fig. 3.4 Structure of processes in the MTS model (Make-to-Stock) (*Source* Own elaboration)

materials, machine production capacity, equipment used in the process of rendering the services)—relative to the prognosed demand for services.

The remaining processes observed for production service providers are similar to those described for product manufacturers. Listed below are the processes and groups thereof characteristic for service providers—Table 3.4, while the structure of processes entailed in rendering production services is presented in Fig. 3.5.

The process of service rendition as such begins with the identification (specification) of the customer's individual requirements, as services tend to have a considerably more dedicated character compared to product manufacturing. As indicated in Table 3.2, services are often rendered parallel to their consumption by the customer. This simultaneity of production and consumption coupled with the direct contact between the producer and the customer result in a situation where the sub-processes of production and service provision merge into a single, homogenous process.

Quality assurance is different for services as compared to products. While in the case of tangible goods it is easy to identify measurable technical parameters, whose satisfaction guaranteed adequate quality of the products, in the case of services qualitative measures are predominant, often based on the subjective opinions of the customer. The quality of services rendered is inherently tied to the customer's attitude and expectations. In turn, service providers are faced with a changeable ability to render services at a specific quality level as the same is susceptible to a variety of factors. Consequently, the systems of quality assurance applied to services tend to utilize very specific methods, e.g. SERWQUAL or Parasuraman's model of quality gaps (Oakland 2014, pp. 101–106).

The reference models characterized below for the production of goods and rendition of services provide a generalization of processes observed in numerous production companies operating under various conditions. The models may serve as a useful tool facilitating the identification and understanding of such processes, as well as a basis for the definition and design of production processes.

Table 3.4 Characteristic of processes/process groups observed in service providers

Process/process group no. (as per Fig. 3.5)	Name of process/process group	Process specifics
1	Design and technical preparation for service provision	The bulk of activities aimed at facilitating the conditions necessary to render services; including development of the service concept as well as structural, technological, and organizational preparation of the company for the requirements of the given service
2	Procurement of materials and utilities	Entails selection of suppliers, procurement of materials and components as well as utilities such as power, gas, and water, including the supply thereof to the company
3	Identification of the customer's specific requirements	Identification or specification of the customer's individual requirements with regard to the service
4	Rendition of services	Performance of activities involved in the rendition of services to the particular customer

(continued)

Table 3.4 (continued)

Process/process group no. (as per Fig. 3.5)	Name of process/process group	Process specifics
5	Service quality assurance	Analysis of the customer's attitude and expectations regarding the service, measures aimed at ensuring the customer's satisfaction
6	Invoicing	Calculating the service charge, issuance of a receipt or invoice, receipt of payment
7	Repairs and maintenance	The processes of repair and maintenance of machines and other technical equipment to guarantee their uninterrupted availability (failure-free and efficient operation) in the course of performing production processes
8	Manufacture/purchase of tooling	Manufacture of tools and instruments for the purposes of own processes, alternatively, purchase of the same from external providers
9	Production strategy	Development of a strategy regarding introduction of new (improved) services, implementation of new technological processes, and development of the production capacity (investments, collaboration)

Process/process group no. (as per Fig. 3.5)	Name of process/process group	Process specifics
10	Planning the demand for production resources	Planning the demand for resources: employees duly prepared to perform particular production tasks, including through adequate training and choice of incentives
11	Production planning and control	The control over service rendition entails planning tasks for the service systems, task queueing, task performance and registering
12	Information systems management	Management of information and ICT systems facilitating the rendition of services
13	Excellence and change management	Management of the development and implementation of improvements with regard to the services and their underlying processes
14	Performance measurement and monitoring	Calculation and observation of indicators illustrating the effectiveness of service processes
15	Discontinuation of production	The decision on discontinuing the provision of a given service and implementation thereof

Source Own elaboration

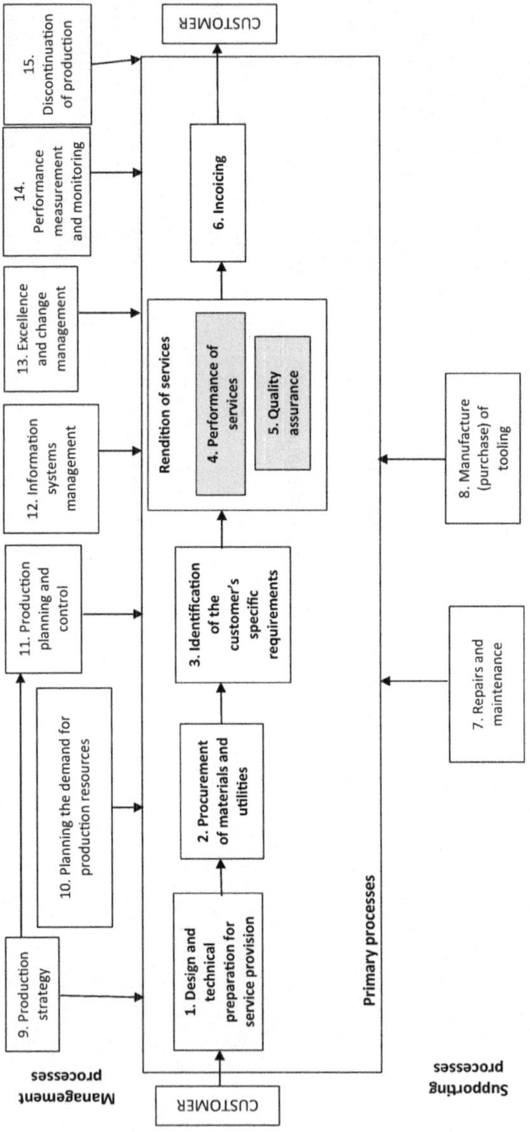

Fig. 3.5 Structure of processes for provision of production services (*Source* Own elaboration)

3.1.3 Process Definition and Design

A characteristic feature of production processes is that a large part of the operations (activities) and input processes involved are fairly specifically defined and designed already at the stage of technical preparations for production. Specific resources are selected (machines, tools, employees) to guarantee adequate quality, lead times are calculated or estimated for the respective operations. Such data are included in documentation compiled at the stage of technical preparation for production and should be used in the course of process design.

In repeatable (stabilized) production, it is not uncommon for detailed process design to be developed already at the stage of technical preparations. Reproducible, standardized products are produced in repeatable, standardized processes that are thoroughly designed down to the tiniest detail. Standardization can pertain to such aspects as machines and equipment, technological parameters, workshop aids, measurement instruments, employee qualifications, work methodologies employed, and even employees' work movements. The process of technical preparation for production is separate from that of production as such, often conducted outside the production facility, e.g. at the manufacturer's dedicated R&D center. The described situation is commonly observed in MTS production.

In non-stabilized, non-repeatable or irregularly repeatable production the level of standardization is naturally lower than in the above case. Products are manufactured upon receiving a customer order (MTO), based on the customer's specifications or choice from the available offer and subject to later customization. The design process is correlated with the production process. Instead of a standardized production, so-called technological routes representative of the respective classes of products are defined. Universal technological resources and broadly qualified employees must be available to perform the operations inherent in such processes.

A defined and designed production process can be illustrated in a model, i.e. a structured process description, usually done in a graphic form. Such models reflect selected key properties of the process, with the defined actual situation ("as is") adopted as the starting point and the desired state of affairs ("to be") as the target.

A variety of computer assisted tools have been employed in recent years for the purposes of process modeling, including UML (Unified Modeling Language) IDEF0, IDEFIX, ARIS Toolset (IDS Prof. Scherr), iGrafx FlowCharter (Corel), or Workflow Analyzer (Meta Software). The BPMN (Business Process Modeling Notation) is currently gaining considerable popularity. There is also a group of more traditional tools dedicated to the modelling of production processes, e.g. (Kanawaty 1992, pp. 89–94):

- outline process chart,
- flow process chart—material type,
- flow process chart—worker type.

as well as the VSM (Value Stream Mapping) method developed by the authors of TPS (Toyota Production System), which served as the foundation for the development of Lean Management (Rother and Shoock 2009).

The tools are aimed not only at reflecting the process "end-to-end" but also include built-in mechanisms to allow identification of value-adding, nonvalue-adding but necessary, and wasteful operations (activities), in line with the value creation classification discussed in Sect. 1.1.2.

3.1.4 Process Redefinition and Redesign

Production processes are identified and defined for the purposes of various systems supporting business activity, including the systems of production planning and quality assurance, IT, and especially ERP (Enterprise Resources Planning). It is therefore important to ensure that the same definitions of processes and process parameters are used across all contexts, which is the role of the standardization prevalent in contemporary production companies. In particular, standardization covers (Liker and Meier 2006, pp. 119–122):

- quality, safety and environmental standards,
- standard specifications: technical information on the correct operation of equipment and certain process specifications required in production,
- standard procedures—used to define operating rules.

The concept of value stream is of key importance to any definition or redefinition of production processes. A value stream is constituted by all the actions (both value added, and non-value added) currently required to bring a product through the main flows essential to every product (Rother and Shoock 2009, p. 3):

- the production flow from raw material into the arms of the customer,
- the design flow from concept to launch.

Redefinition and redesign of production processes is based on an analysis of the value stream relative to the respective processes and identifying possibilities of reducing nonvalue-adding but necessary and wasteful operations. This topic will be discussed in detail in Sect. 6.1. Improvement of production processes.

3.2 Process Modeling in Public Administration—Katarzyna Rostek

3.2.1 The Context of Public Administration

The economic and political crises of the 1980s and 1990s, weaknesses of the overinflated public sector, and criticism of the traditional model of administration, inspired the search for more effective and efficient forms of governance. Improvement of its quality was perceived in the context of adapting market mechanisms to the needs as well as managerial methods and techniques broadly used in business to the needs of the public sector. Efforts were also made to shift the focus of administration towards better operating performance, greater economic efficiency, quality, and overall results. The reform initiated with a view to achieving those results is known as New Public Management (NPM) (Hood and Jackson 1992).

The concept of NPM revolves around gaining a better understanding of the needs of customers (citizens) to whom public services are addressed, and ongoing identification of changes in terms of legal requirements and regulations. It is introduced through activities aimed at generating public value by rendering high quality public services (Fig. 3.6).

NPM emphasizes:

- correct identification of goals and monitoring results, financial management—effectiveness of services,
- definition and monitoring of standards—repeatability and stable quality of services,
- utilization of benchmarking studies—high quality of services,
- modern methods of HR management—productivity of services.

The same can also be described in the form of a triad (Ferlie 2017): Market—Measurement—Management (3M), where the market is understood as the general environment wherein public services are provided, which ought to be managed in such a way so as to achieve the

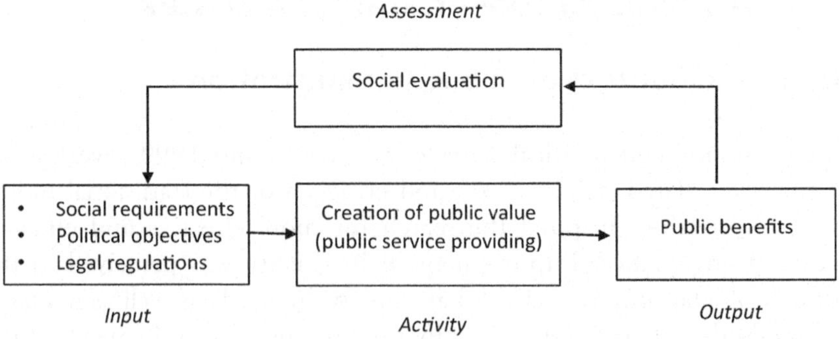

Fig. 3.6 Public value in New Public Management (*Source* Own elaboration)

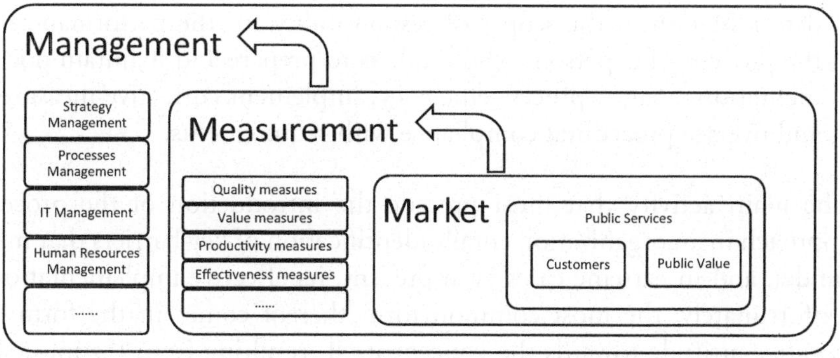

Fig. 3.7 Market-Measurement-Management triad (*Source* Own elaboration)

highest values of quality and efficiency measures relative to the public value created (Fig. 3.7).

Actual application of the 3M triad with the aim of ensuring the creation of public value at a level consistent with social expectations, requires the implementation of the process approach in the organization, i.e. (Kauf 2015, pp. 127–128):

- precise identification of external factors and circumstances—social, political, economic, and legal—that condition the fulfilment of public roles by the organization,
- adequate definition of goals—which are of key importance from the perspective of performing public tasks,
- apt identification of processes, including:
  - strategic (so-called mega-processes)—which are crucial to socioeconomic development (e.g. resulting from the implications of the adopted social policy of economic growth),
  - primary—related to the efforts directly related to the generation of public value (e.g. imposed by the provisions of law),
  - supporting—facilitating efficient and correct performance of primary processes in terms of creating public value,

- determination of the scopes of responsibility for the performance of the process—i.e. persons whose role is to prepare and maintain documentation, analyze process efficiency, implement corrective measures, and oversee procedural compliance with requirements.

The main activity that must precede the introduction of the process approach in an organization entails identification of key barriers that may hinder, and in extreme cases even prevent, its effective implementation. Unfortunately, the most common form thereof comes in the form of negative attitude towards the concept itself, resulting from (Szumowski 2009, p. 110):

- a negative perception of the methods and tools associated with the process approach,
- the organization's monopolistic position,
- insistence on the possibility of continued effective operation in the functional mode,
- belief that provisions of law sufficiently define the organization's operations,
- insistence that the quality of services is not the primary objective of the organization,
- the necessity to generate additional documentation perceived as redundant by the employees,
- small size and limited complexity of the organization, serving few customers.

Overcoming said barriers is the necessary first step when electing to implement the process approach. Subsequently, one must also tackle actual limitations faced by entities aiming to do so (Szumowski 2009, p. 111):

- shortage of qualified personnel, also personnel shortages throughout the organization,
- likely personnel resistance to the change,

- lack of confidence in the need for innovation among managerial staff, also lack of tangible support from the management for the introduction of the process approach,
- expected high cost of implementation couples with financial difficulties faced by the organization,
- lacking stability of authority in the organization, as well as excessive hierarchization and formalization of the power structure (which enforces preservation of strictly defined decision-making chains),
- disfunctions existing in the current operating structure that would be revealed and eliminated should the process approach be implemented,
- other priorities that outweigh and delay the transition to processually-centered operations.

As follows from the above, every attempt to implement the process approach in public administration will be an organizationally, financially, and politically complex decision requiring comprehensive preparation and reliable approach. A number of important factors and suggestions applicable in this regard will be discussed in the subsequent chapters of this study.

3.2.2 Process Identification

The primary goal of identification efforts is to develop a specific ecosystem (Fig. 3.8), whose foundation is grounded in law and cross-institutional processes that condition the performance of respective organizational goals. Those goals are later elaborated to account for specific services and tasks in such a way, so as to maximize compatibility with customer needs and ensure adequate cross-institutional cooperation.

The diagram presented in Fig. 3.8 has been practically and successfully verified in the context of Swiss public administration, and the concept of process management introduced therein was effectively developed and implemented based on this exact approach (Ahrend et al. 2012, pp. 8–10). The basis for process identification is provided by the applicable laws and ongoing cross-institutional cooperation related, for instance, with obligations to provide public information, provision of outsourcing

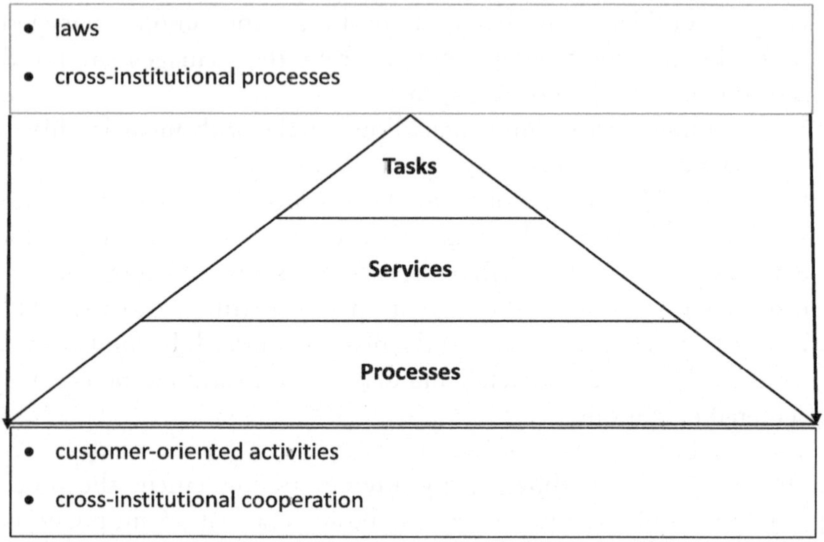

Fig. 3.8 Ecosystem of administrative processes management (Based on Ahrend et al. 2012, p. 9)

services, or coordination of inter-institutional operations. Said elements define specific goals that must be realized by any public administration institution under given conditions and within a given timeframe. The performance of such tasks relates to the structuration and rendition of public services, which in the institutional context constitutes a process or group of processes applicable to the given organization. As a result, the customer receives a comprehensive benefit (a public service is rendered), and the cooperating institutions are provided with results of the tasks assigned.

This approach is logical and viable in the context of public administration but does not immediately solve every problem. As evidenced by the Swiss example, implementation of the process approach was a very slow process, the reasons for which were primarily associated with low personnel awareness, particularly with regard to management, of the attainable effects and benefits thereof, which in turn translated to insufficient support for the project as such from the managerial staff. As mentioned above, this barrier ought to have been eliminated before initiating the transition as its activation already during the implementation significantly delayed and hindered the achievement of expected results.

3.2.3 Process Definition

Given the existing barriers and limitations coupled with the simultaneous need to support the concept of NPM, it is evident that a dedicated approach to process definition and design must be developed for public administration, one that will duly account for its character and specificity. For although the definition of process as such remains the same in the context of both economic activity and public administration, the actual specific processes vary considerably between the two, particularly in terms of e.g. capacity for standardization, stability in time, or susceptibility to change. This difference is reflected in one of the currently popular classifications of administrative processes cited in Sect. 1.1.2 (Ziemba and Obłąk 2014, p. 625) which divides processes into the categories of routine, negotiable, and poorly structured.

Particularly in the case of non-routine processes, it is helpful if process identification and definition follows the sequence of systematized and standardized activities with the use of specialized IT tools and adequate substantive support. One example of such support is provided by the Seven Process Modeling Guidelines (7PMG) which provide the basic rules facilitating the viability and effectiveness of developed process models (Mendling et al. 2010, p. 130):

- G1: Use as few elements in the model as possible,
- G2: Minimize the routing paths per element,
- G3: Use one start and one end event,
- G4: Model as structured as possible,
- G5: Avoid OR routing elements,
- G6: Use verb-object activity labels,
- G7: Decompose the model if it has more than 50 elements.

7PMG guidelines have also been expanded in the form of the Ten Process Modeling Rules (Mendling 2012, p. 104), wherein rules 4 (model structure) and 5 (model simplicity) were further elaborated to include the following sub-items:

- G1 Nodes: Do not use more than 31.
- G2 Connectors Degree: No more than 3 inputs or outputs per connector.
- G3 Start and End: Use no more than 2 start and end events.
- G4.a Structuredness: Model as structured as possible.
- G4.b Mismatch: Use design patterns to avoid mismatch.
- G5.a OR-connectors: Avoid OR-joins and OR-splits.
- G5.b Heterogeneity: Minimize the heterogeneity of connector types.
- G5.c Token Split: Minimize the level of concurrency.
- G6 Text: Use verb-object activity labels.
- G7 Nodes: Decompose a model with more than 31 nodes.

The use of specialized IT tools requires the application of implemented and familiar notations and techniques. The description of processes performed in public administration may be done with the use of standard tools. Currently, the most prevalent process modelling standard is BPMN 2.0 (Business Process Model and Notation),[3] supported by many providers of IT solutions. Other tools include EPC (Event-Driven Process Chain)[4] diagrams based on ARIS architecture, as well as the BPMS (Business Process Management System)[5] method, developed on the basis of ADONIS software. One could also consider applying UML (Unified Modeling Language)[6] diagrams, particularly if the process approach is implemented alongside supporting IT solutions.

Methods of automated or semiautomated process discovery also exist, which utilize algorithms (like the genetic or ant algorithm) that generate

[3]https://www.omg.org/spec/BPMN/2.0/PDF, read 2020-07-13.

[4]https://www.ariscommunity.com/event-driven-process-chain, read 2020-07-13.

[5]https://fr.boc-group.com/uploads/files/_2015__ADONIS_Whitepaper_-_EN_04.pdf, read 2020-07-13.

[6]https://www.omg.org/spec/UML/2.5.1/PDF, read 2020-07-13.

optimized process variants based on a reference model thereof. These can be successfully utilized in subsequent iterations of the lifecycle, but their effectiveness will nonetheless strictly depend on the precision, detail, and adequacy of definitions and structure of the base model.

3.2.4 Process Design

Regardless of the type of process being designed (routine, negotiable, case-by-case, or poorly structured), one needs to consider a wide spectrum of factors that affect and determine its correct performance. Hence, solutions are needed that would account for the specificity of public administration processes. An example of such a dedicated solution is the concept of Government Process Management (GPM), which means that rules and practices derived from the BPM concept are applied but in a way adapted to the specificity of public administration rather than economic organizations (Tregear and Jenkins 2007, p. 2). The concept is based on identification of barriers and differences observed between process management in the context of commercial companies and public administration institutions, and subsequently appropriate adaptation of the BPM approach to increase the likelihood of its successful implementation and use. The mentioned differences pertain to a very broad spectrum of aspects, including e.g. (Syed et al. 2018, pp. 3–5): mission and vision, aims, process customers, process requirements, public accountability, economic structure, political sensitivity, ecosystems, budget cycle complexities, information exchange, regulatory society, machinery of government change, and culture.

Upon an analysis of the above characteristics, it can be observed that they can be grouped into three distinct categories (Cherouana et al. 2019, p. 70): economic-legal (e.g. public accountability, economic structure, political sensitivity, ecosystems, regulatory society, machinery of government change), organizational (e.g. mission and vision, aims, information exchange, budget cycle complexities, culture), and domain (e.g. process customers, process requirements). Each of the categories constitutes a separate scope in the design of administrative processes (Fig. 3.9) relative to the factors affecting the same.

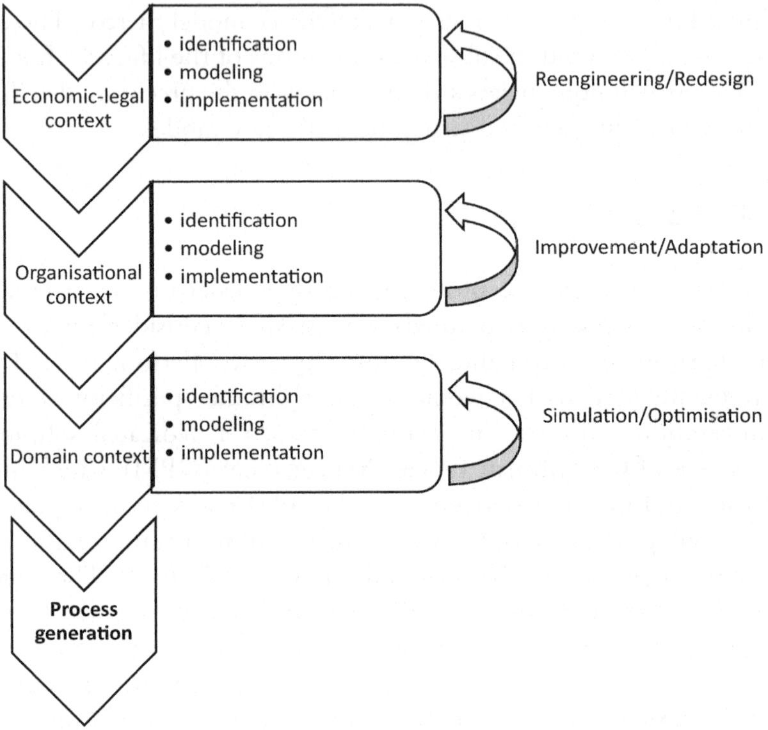

Fig. 3.9 Context approach to the generation of public sector processes (*Source* Own elaboration)

As follows from Fig. 3.9, the stage of modeling in the context of economic-legal factors ought to precede modeling in the context of organizational and domain factors in such a way, so as to ensure possibility of securing and satisfying compliance with all organizational requirements, social expectations, and legal/economic limitations. At the same time, further improvement and changes in terms of administrative processes ought to also take the distinction between these two perspectives into account. As evidenced in Fig. 3.9, the context of given changes will also induce their scope, hence the most far-reaching and significant changes with respect to public sector processes occur in the context of economic and legal factors. In the organizational context, the existing processes are subject to correction and improvement, while at the domain level, individual processes undergo local optimization, for instance in terms of more efficient resource management.

The catalogue of public administration processes generated under this approach constitutes an important element of cross-institutional and social communication. Therefore, the processes ought to be modeled in accordance with the rules stipulated in the Organization Transparency Catalogue, where a five-step model is proposed to define the organizational transparency concept (Cappelli and Leite 2010):

- accessibility—information about the organization is available to the external environment,
- usability—available information can be easily obtained and used,
- informative—information is made available with expected quality,
- understandability—external users can understand the available information,
- auditability—external users can manipulate, comment and even generate new data and information on the basis of available information.

In order to account for a deeper interpretation of the technical context of process modeling, the above characteristics should be elaborated as follows (Engiel et al. 2014, p. 97):

- adaptability—the ability to change a process and its representation in accordance with stakeholders' needs,
- clarity—the ability to unambiguously identify any object/element used in process representation,
- conciseness—the ability to summarize content to facilitate faster information availability and exchange,
- intuitiveness—the ability to allow stakeholders to read process model representations without previous knowledge of the domain and/or the notation used to describe the process,
- simplicity—the ability to use less technical language and make use of well-known metaphors/analogies to represent process elements,
- uniformity—the ability to provide a unique description and representation for each process element.

The presented understandability attributes of public service process models often become the basis for objective process redefinition in an organization. The Polish ZUS (Social Insurance Institution) may serve as a good example here. The project of general process reorganization implemented by the institution (Rostek et al. 2018, pp. 125–126) focused in particular on clarity, conciseness and simplicity, which resulted in a significant reduction in the number of processes (from 462 to 41), better understandability thereof for the customers and employees alike (simplified language used in definitions and communication), and consequently higher quality of service and improved perception and credibility of the institution as such.

3.2.5 Project Redefinition and Redesign

Naturally, not every context or process within an organization will allow a precise (routine) definition. Nonetheless, each one still needs to be identified and described to the extent of its possible customization (through modification, adaptation to a given situation, or negotiation). Notably the limits of said extent must be controlled and updated to account for changes in applicable legal regulations, social expectations, or economic and political circumstances. It is therefore extremely important for the identification and differentiation of processes to account for all the key elements that may influence the correctness and effectiveness of their performance. One way to assess this is to use the context of the developed model's quality. An example of a tool applicable for this purpose is the SEQUAL model, which covers the following evaluation criteria (Krogstie 2016, pp. 83–88):

- physical quality—the model must be sufficiently persistent, current, and available relative to the goals of modeling and those involved,
- empirical quality—the model has to be usable in the context of communication between people,
- syntactic quality—all statements in the model are according to the syntax and vocabulary of the modeling language or notation,

- semantic quality—the language/notation used makes it possible to represent the necessary processual aspects within the given domain (domain appropriateness),
- pragmatic quality—comprehensiveness of the model for participants considered in two dimensions: how the model is interpreted by human stakeholders and what it is intended to express,
- social quality—compatible interpretations by different social actors,
- deontic quality—improvement of the business process in terms of time, quality, cost, flexibility, resource consumption, and service level.

A model is defined as a simplified description of a system or process to assist calculations and predictions. And if it is to serve as a simplified representation of reality, it ought to include only the key elements that comprise the same, rather than all of them. Hence, defining base criteria for the assessment is pivotal to the reliability and usability of its results. Such criteria will then serve as the basis for the generation of improved versions of the model in the subsequent iterations of the lifecycle. New model variants should be assessed in five dimensions (Buijs et al. 2012, p. 45):

- replay fitness—the dimension points to how the discovered model can accurately replay the cases recognized through previous process iterations,
- precision—the dimension shows if the discovered model prohibits action which is not seen in the reference model,
- generalization—the dimension assesses the extent to which the resulting model will be able to reproduce possible future, yet unseen, process activities,
- simplicity—according to the Minimal Description Length (MDL) paradigm (Grünwald 2007), the model quality is no longer only based on predicting performance, but also on the simplicity of the model,
- similarity—a new version of the reference model can be found that is similar to the initial reference model yet is improved with respect to its fit with actual behavior.

The conducted assessment will serve as the basis in the definition (automated or manual) of a new reference model, and consequently a new process standard. In public administration institutions, which are characterized by elaborate structures and high levels of autonomy of respective organizational units, the establishment of new standards will require numerous mutual arrangements to be agreed upon between such units as well as, oftentimes, even key process operators. In such a case, any projected change (redefinition or redesign) will require a team to be established with the participation of all primary decision-makers responsible for the given process. It may also be advisable to establish a separate unit or team within the organizational structure, whose members will be responsible for ongoing management of projects concerned with altering processes with a view to selecting those likely to most effectively facilitate the achievement of the strategic goals of the entire organization, rather than individual representatives thereof. Such teams are particularly viable in large organizations such as state administration, ministries, or universities.

3.3 Comparison and Recommendations—Anna Kosieradzka and Katarzyna Rostek

The comparison between the perspectives of production companies and public administration institutions will be divided into thematic sub-chapters to better characterize the key distinguishing factors and similarities between the two groups.

3.3.1 Organizational Context

There are certain fundamental differences distinguishing the application of the process approach in production companies and public administration institutions. In production organizations, every process is a natural consequence of operations that focus on and are rooted in production serving as the core around which all remaining processes, procedures, and structures are developed with a view to facilitating and managing production. Hence, in order to function efficiently, a production organization must rely on process management, since processes constitute the most fundamental and natural form of its effective operation.

The situation is quite different in the case of public administration. In this context, the notion of processual approach was only introduced in an effort to seek more effective methods and improvements to the performance of such organizations and the public value they generate. Since primary problems stemmed from the functional approach to management, process management was suggested as a possible method of achieving beneficial changes. However, many problems were encountered in this context that have been and likely will be for a long time to come the focus of research scrutiny by both theoreticians and practitioners. Said problems stem from the fact that processes, their organization, significance, and classification are distinctly different from what one can encounter in production companies.

One should therefore begin by emphasizing that process management in the two analyzed types of organizations will take on different forms, will often utilize different methods and tools, and will even aim to achieve differently formulated goals, which does not mean, however, that specific models and experiences cannot be shared between the two contexts. For this reason, the identification of similarities and differences will integrate our further deliberations, in particular in terms of the conclusions in subsequent chapters.

3.3.2 Process Identification

In a commercial organization manufacturing products or rendering services, the applicable process map is well known and virtually predefined. Naturally, certain differences could be identified depending on the type of the organization, the methods and scope of its activity, extent of the output offer, etc. Nonetheless, we will still be operating within well-known limits and the structures and schemes already discussed in Sect. 3.1.2. Meanwhile, there is no such uniformity in public administration because the applicable process maps stem from three diverse variables: provisions of law, cross-institutional correlations, and public utility. Each of these elements has a distinct bearing on the eventual shape of organizational processes, which means that they tend to be highly unique and customized.

It is also noteworthy that while identification of processes in the case of production organizations is comprehensively supported by standardized procedures and codes of conduct, in the case of public administration we have only general guidelines to work with, with specify overall good practices but need to be individually adapted to the given context. Due to the above, such projects tend to be highly unconventional and original, and therefore difficult to implement and burdened with the risk of failure to achieve the expected results.

The ability to reduce the severity of said risk even before initiating project activities is often a factor that determines the success of the entire initiative. Measures that allow one to do so include thorough preparation of the organization for the implementation of the process approach (e.g. by increasing employee awareness, securing support for the project, ensuring the availability of resources at the right time and adequate level), guaranteeing the necessary competence and experience of the project team, but also increasing the organization's general openness to changes that are bound to result from the introduction of the process approach. Without ensuring the appropriate conditions at the onset of the project, one cannot hope to effectively proceed with its implementation, not to mention hope for its successful conclusion.

3.3.3 Project Definition and Design

Project definition and design in production companies is a highly normalized and standardized endeavor. Process designs are formulated in accordance with the overriding goal of manufacturing products for the customer in line with the quantity, quality, and time requirements adopted. A process design is then verified and approved based on its capacity to meet specifically formulated requirements and satisfy specific quality and efficiency criteria. Such activities are supported by a broad spectrum of tools, methods, and techniques derived from BPM.

In the case of process definition and design in public administration institutions, the same tools, methods, and techniques can also be employed. This is due to the fact that the primary differences between the discussed perspectives are legal, organizational, and functional in nature, rather than technical or methodological. Notably, while the additional requirements regarding legal aspects, cross- and inter-institutional cooperation, and public utility have led to the proposed replacement of BPM with GPM, the latter continues to rely on standard notations and techniques that tend to apply well regardless of the type of organization.

The significance of designed processes in the context of the organization's environment must also not be overlooked. In the case of production companies, the same constitute proprietary assets and contribute to the organization's overall value. Such organizations will therefore hardly be interested in making the same public knowledge as the processes are treated as elements of intellectual capital. The only aspects that are shared outside are those encountered at meeting points, usually logistic connections, sometimes also in the context of cooperative engagement with external entities. Meanwhile, in the case of public administration, processes are designed with a view to achieving social goals and creating public value. They therefore tend to be transparent elements of organizational management that can be shared and discussed with external stakeholders.

3.3.4 Process Redefinition and Redesign

In production organizations, the redefinition and redesign of processes is a permanent element of control and optimization activities. It is also aimed at continuous improvement of efficiency in terms of resource consumption and consequently maximization of profits. In governmental administration, the search for the best, i.e. most current, adequate, and compatible configuration of processes is still ongoing. This results in frequent changes and redefinitions observed not only in the context of the structural definition of individual processes but also entire process maps of respective institutions. Consequently, any such endeavors undertaken by public administration entities tend to take the form of in-depth and often rather difficult organizational changes, while in production companies they usually only pertain to individual processes and elements conditioning their execution.

The process map applicable to a production organization is stable and fixed in time through the duration of its economic activity. Meanwhile, a corresponding map in a public administration institution will undergo changes whenever a comprehensive reorganization of its processes is implemented or the applicable legal regulations or assigned public goals are modified. This results in a situation where the levels of uncertainty and risk associated with process modeling in public administration are significantly higher than those observed in production companies. In turn, the stability and repeatability of processes observed in production organizations facilitates their standardization and more effective selectin of supporting tools. In public administration we are faced with uniqueness and high complexity of every situation entailing process modeling. It is therefore vital that methods, tools, and techniques derived from the context of production companies be duly adapted to each respective case so that they can provide effective support in the highly uncertain context of public administration institutions. Naturally, public administration will, in time, develop a canon of dedicated methods, tools, and techniques. Until then, however, it should take advantage of the experience and knowledge of organizations whose resources mat significantly contribute to their development.

References

Ahrend, N., Walser, K., & Leopold, H. (2012). Comparative Analysis of the Implementation of Business Process Management in Public Administration in Germany and Switzerland. In K. Sandkuhl, U. Seigerroth, & J. Stirna (Eds.), *Emerging Topics in the Practise of Enterprise Modeling* (pp. 1–12). 5th IFIP WG 8.1 Working Conference, PoEM 2012.

Buijs, J. C., La Rosa, M., Reijers, H. A., van Dongen, B. F., & van der Aalst, W. M. (2012, June). Improving Business Process Models Using Observed

Behavior. In *International Symposium on Data-Driven Process Discovery and Analysis* (pp. 44–59). Berlin and Heidelberg: Springer.

Cappelli, C., & Leite, J. C. S. P. (2010). Software Transparency. *Business and Information Systems Engineering, 2,* 127–139.

Cherouana, A., Mahdaoui, L., Bellatreche, L., & Medjahed, B. (2019). A Semantic Approach for Generating Government Processes. *International Journal of Web and Grid Services, 15*(1), 59–92.

Engiel, P., Araujo, R., & Cappelli, C. (2014). Designing Public Service Process Models for Understandability. *Electronic Journal of e-Government, 12*(1), 95–111.

Ferlie, E. (2017). The New Public Management and Public Management Studies. *Oxford Research Encyclopedia of Business and Management*. https://doi.org/10.1093/acrefore/9780190224851.013.129.

Grünwald, P. D. (2007). *The Minimum Description Length Principle.* Cambridge: MIT Press.

Heizer, J., & Render, B. (2011). *Operations Management* (10th ed.). Upper Saddle River, NJ: Pearson.

Hood, C., & Jackson, M. (1992). The New Public Management: A Recipe for Disaster? *Hazard Management and Emergency Planning: Perspectives on Britain* (pp. 109–125). London: Routledge.

Kanawaty, G. (1992). *Introduction to Work Study*. Geneva: ILO.

Kauf, S. (2015). Orientacja procesowa w zarządzaniu publicznym łańcuchem dostaw. *Studia Ekonomiczne, 251,* 122–131.

Krogstie, J. (2016). Quality of Business Process Models. In *Quality in Business Process Modeling* (pp. 53–102). Cham: Springer.

Liker, J. K., & Meier, D. (2006). *Toyota Way Fieldbook*. New York: McGraw-Hill.

Melnik, S. A., & Denzler, D. R. (1996). *Operations Management: A Value-Driven Approach*. Chicago: Irwin.

Mendling, J. (2012). Managing Structural and Textual Quality of Business Process Models. In *International Symposium on Data-Driven Process Discovery and Analysis* (pp. 100–111). Berlin, Heidelberg: Springer.

Mendling, J., Reijers, H. A., & van der Aalst, W. M. (2010). Seven Process Modeling Guidelines (7PMG). *Information and Software Technology, 52*(2), 127–136.

Oakland, J. S. (2014). *Total Quality Management and Operational Excellence.* New York: Routledge.

Rostek, K., Wiśniewski, M., & Gołuch-Trojanek, K. (2018). Uwarunkowania procesu wdrażania zintegrowanych systemów informatycznych w instytucjach publicznych. *Nowoczesne Systemy Zarządzania, 13*(4), 117–129.

Rother, M., & Shoock, J. (2009). *Learning to See: Value Stream Mapping to Create Value and Eliminate Muda.* Brookline: Lean Enterprise Institute.

Syed, R., Bandara, W., French, E., & Stewart, G. (2018). Getting It Right! Critical Success Factors of BPM in the Public Sector: A Systematic Literature Review. *Australasian Journal of Information Systems, 22,* 1–39.

Szumowski, W. (2009). Zastosowanie podejścia procesowego w urzędach administracji samorządowej. *Prace Naukowe Uniwersytetu Ekonomicznego we Wrocławiu,* no. 52 Podejście procesowe w organizacjach, pp. 106–118.

Tregear, R., & Jenkins, T. (2007). *Government Process Management: A Review of Key Differences Between the Public and Private Sectors and Their Influence on the Achievement of Public Sector Process Management.*

Ziemba, E., & Obłąk, I. (2014). Informatyczne wsparcie procesów w administracji publicznej. *Roczniki Kolegium Analiz Ekonomicznych/Szkoła Główna Handlowa, 33,* 619–646.

4

Process Implementation

Dr. Anna Kosieradzka and Katarzyna Rostek

4.1 Process Implementation in Production Companies—Anna Kosieradzka

In production companies, processes are introduced by way of implementing a particular concept of production management developed on the basis of the process approach.

4.1.1 Implementation Strategy

Table 1.5 in Chapter 1 characterized the popular management concepts used in production and rooted in the process approach. A management concept is understood as a cohesive and comprehensive approach, complete with a set of rules and guidelines as well as a range of methods and tools developed for the purposes of the given approach or derived from its earlier predecessors. Concepts such as Lean Management, Theory of Constraints, and TQM are commonly employed in the implementation of the process approach in organizations, whereas

A. Kosieradzka and K. Rostek, *Process Management and Organizational Process Maturity*,
https://doi.org/10.1007/978-3-030-66800-6_4

Business Process Re-engineering, Kaizen, and Six Sigma are focused on process improvement and will be discussed in Chapter 5.1.

Conditions conducive to the implementation of the process approach in an organization are established by the adoption of the TQM concept as the leading management approach. The basic principles of TQM were created by American "gurus" of quality management such as W.E. Deming, J.M. Juran, P.B. Crosby, and A.V. Feigenbaum. Deming is additionally considered to be one of the main contributors to Japan's economic success in the 1960s and 1970s. The works of those authors led to the formulation of the universal quality management guidelines included in the ISO 9004:2000 standard: *Quality management systems—Guidelines for performance improvements* (revised by ISO 9004:2009 and ISO 9004:2018), where the process approach is strongly accentuated. The leading principles included therein are customer orientation [1], leadership [2], and engagement [3] of all employees in quality-related decision making. An organization is perceived as a system (systemic approach [5]) wherein customer value is generated by processes [4] – see Fig. 3.1). The systemic approach facilitates a better understanding of the mutual interactions between processes. Decisions are made based on facts [7], and continuous improvement [6] of processes and products requires the application of adequate methods. The picture is completed by partnership with suppliers [8] that determines the ability to meet the requirements of the end customer due to the close co-dependence between entrepreneurs and their suppliers.

J.S. Oakland (2014, pp. 27–30) posited that the TQM comprises "four Ps and three Cs":

- *Planning* – the development and deployment of polices and strategies,
- *Performance* – measuring framework, self-assessment, audits, reviews and benchmarking,
- *Processes* – understanding, management, design and redesign, continuous improvement,
- *People* – managing human resources, with due account for *Culture*, *Communication* and *Commitment*.

Lean Management (LM) is a concept that advocates a comprehensive approach to the synthetic implementation of processes in production companies and constitutes an elaboration of solutions developed under TPS (Toyota Production System). The term "Lean Production" was first used and defined as part of the famous research project known as IMVP (International Motor Vehicle Program) and conducted at MIT based on studies conducted in the context of automotive industry companies from around the world—primarily Japan, the United States, and Germany. Currently, Lean Management has gained the status of a universal concept that can be applied to every type of economic activity. It basically perceive the use of resources for any purpose other than creating value for the customer as wasteful, in line with the core concept of the process approach. The implementation of lean production tends to yield the best results in companies engaged in repeatable, large-scale production of standardized products, where every, even the smallest saving in resource consumption can translate to considerable improvement at the scale of the entire enterprise. Nonetheless, the principles of LM are universal enough to be applicable to any business entity. Womack and Jones (1996, pp. 16–28) referred to the same as Lean Thinking and elevated the same to the status of a business philosophy:

- Specify Value – value can only be defined by the ultimate customer,
- Identify the value stream – the value stream is a set of all specific actions required to bring a specific product in the hands of customer,
- Flow – once value has been precisely specified, value stream fully mapped and obviously wasteful steps eliminated, it is time to make the remaining, value-creating steps flow,
- Pull – as flow is introduced, let customers pull value from the next upstream activity,
- Perfection – accomplishing Steps 1–4 is a great start, but the fifth step is perhaps the most important: making lean thinking and process improvement part of the corporate culture.

Somewhat overshadowed by the lean concepts that eventually became the standard in word-class manufacturing, the Theory of Constraints (TOC) was proposed by an Israeli physicist, Eliyahu M. Goldratt, in the 1980s – i.e. almost parallel to the emergence and development of the Lean Production concept. Goldratt's proposal can be said to stem from his observation that (Cox and Schleier 2010, p. 3) "*Focusing on everything is synonymous with not focusing on anything*". Rather than perfect all elements of the production system, one ought to focus on the constraints responsible for the emergence of process bottlenecks which reduce production flows and generate losses. Every organization has only a few constraints, elimination of which leads to radical improvement of results.

The following principles of the Theory of Constraints can be formulated (Cox and Schleier (Ed.) 2010, pp. 906–915):

- Step 1 – Identify the system constraint,
- Step 2 – Decide how to exploit the constraint,
- Step 3 – Subordinate everything else to the above decision, including other resources and activities,
- Step 4 – Elevate the system's constraint,
- Step 5 – Go back to Step 1 (the same procedure is repeated for a new constraint).

The principles of TOC are also fairly universal and can be applied to the introduction of the process approach in a variety of organization types.

Under both LM and TOC, dedicated tools were developed (methods and techniques) to facilitate the practical implementation of the respective rules, and thus to serve the purposes of introducing the process approach in a company. TQM and LM take advantage of a shared repository of tools, whereas the Theory of Constraints relies on DBR (Drum-Buffer-Rope)—an application dedicated to production companies, while also not shying away from deriving solutions from the TQM and LM toolset.

4.1.2 Process Implementation Methods

As already stressed above, implementation of the process approach in production companies takes the form of a targeted introduction of partial methods consistent with the principles of two concepts. The main tools developed in the context of LM are presented in Table 4.1, while the TOC toolset is characterized in a Table 4.2.

Value stream is a key LM concept and VSM (Value Stream Mapping) plays a vital role among process implementation methods proposed by Lean Management. The method allows one to identify material and information flows in the production process, and to understand exactly where and how customer value is created. It also facilitates the selection of adequate methods with a view to eliminating losses that emerge in the process and limiting activities that do not directly contribute to the creation of customer value but only support the same. A set of standardized symbols is commonly used when preparing such maps. The guidelines applicable to the development and utilization of Value Stream Maps were discussed by Rother and Harris (2001), Rother and Shoock (2009), and Jones and Womack (2011).

Figure 4.1 A graphic interpretation of the basic TOC DBR tools discussed in Table 4.2.

Table 4.1 Tools utilized in Lean Management

Principle	Tools
1. Value	Value Engineering, unification, Group Technology, structural modularity, management of component diversity
2. Value stream	VSM – Value Stream Mapping, identification of VA (Value adding), NVA (Non-value adding but necessary), and Wasteful (*muda*) activities; waste elimination,
3. Flow	One-piece flow, Small batch-size, Autonomation, Heijunka (leveling of production), SMED – quick changeover; TPM (Total Productive Maintenance), SPC (Statistical Process Control)
4. Pull	Kanban, Supermarket, JIT (Just-in-Time), Partnership with Suppliers
5. Perfection	5S, Poka-Yoke, Visual Control, Self-inspection, Deming's PDCA Cycle, Continuous Improvement - Kaizen, work standardization - SOP (Standard Operation Procedures), Quality Circles, team problem solving - SGA (Small Group Activity), multitasking skills, Competency Matrix, employee training and development.

Source Own elaboration

Table 4.2 Tools utilized in TOC

Principle	Tools developed under TOC	Tools derived from LM
1. Identifying the constraint	Identifying the constraint is the starting point, because the constraint affects the size of the production output. A constraint is defined as the heaviest burdened resource before which tasks awaiting execution are accumulated.	VSM
2. Exploiting the constraint	This step utilizes all available methods of increasing the production flow through the bottleneck, e.g.: relieving the bottlenecked resource of all tasks that may feasibly be executed elsewhere, minimizing the time needed to retool the bottleneck, prioritizing all relevant technical maintenance, guaranteeing that no inferior quality materials are introduced into the bottleneck, etc. The result of all these activities is **the drum**, i.e. a production schedule maximizing the use of the bottleneck capacity.	SMED – quick setup TPM – Total Productive Maintenance Poka-Yoke 5S

Principle	Tools developed under TOC	Tools derived from LM
3. Subordinating all resources and activities to the constraint	**The buffer** and **the rope** allow for almost every process in the enterprise related to production planning, materials purchase, or shipment of finished products to be subordinated to the constraint. The buffer protects the bottleneck from random process fluctuations that generate deviations from the schedule. The rope is a mechanism used in identifying the exact moment of introducing the material to the first operation of the production process so that the right amount of intermediate products reaches the bottleneck in time.	Kanban
4. Elevating the constraint	**Elevating** the constraint should be considered only after exploiting all the potential of the constraint. When there are still possibilities of increasing sales (market is not a constraint), elevating the constraint can be done through increasing its throughput, e.g. by buying a new machine.	SOP
5. Return to step 1	Returning to step one completes the cycle of continuous development – the same procedure is then repeated for a new constraint.	Kaizen

Source Own elaboration

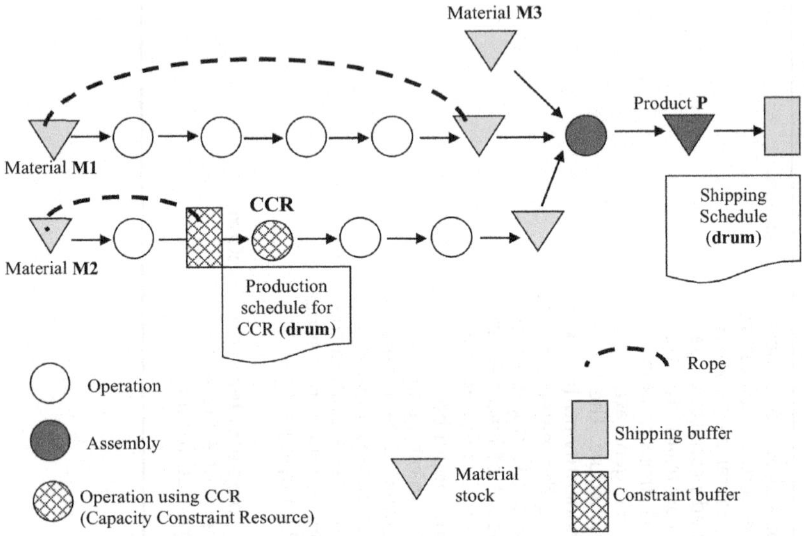

Fig. 4.1 Interpretation of TOC Drum-Buffer-Rope tools

The main differences between the LM and TOC concepts are highlighted in Table 4.3.

There are a number of conclusions that can be derived from the above juxtaposition:

- LM is a system dedicated to repeatable production against fairly stable demand, product unification, and process standardization. Lean Manufacturing may be supported by such TOC elements as:
  - identification of bottlenecks in the value stream,
  - focus on eliminating losses and wastages, however, not in the entire system but rather in areas that may directly contribute to improved performance,
- In the case of non-stabilized, small to medium-scale production, the Theory of Constraints ought to be the primary concept in the implementation of the process approach, with support from such LM elements as:

Table 4.3 Comparison of LM and TOC concepts

Trait	LM	TOC
Goal	Cost reduction	Throughput improvement
Means of achieving the goal	Reduction of all wastage	Only reduction of wastage related to the constraint matters
Role of the constraint	All resources are equally important	Time wasted on the constraint translates to reduced throughput of the entire system.
Measures	Improvements can be introduced based on local productivity measures	Consistency of local and global measures means that local optimization cannot be accounted for in the decision-making process
Variability	Efforts aimed at eliminating all variability: internally through work standardization and externally by more flexible organization of production	Acceptance of internal and external (market) variability as a fact of life and protection against its consequences through buffer management
Burden on resources	Even – balance load on all resources	Utilization of resources other than bottlenecks is of marginal importance
Batch size	Constant for all links of the process	Simultaneously adapted to the respective phases of the process
Planning standards	Production flow rhythm Production cycle	None. The approximate duration of the cycle depends on current schedules
Employee participation	Subjective – team-based problem solving and implementation of proposed improvements	Objective – implementation of improvements by overcoming potential resistance from employees

Source Own elaboration

- value stream mapping,
- flow,
- SMED,
- TPM,
- Kanban,
- 5S,
- Poka-Yoke and quality tools.

A hybrid LM and TOC approach needs to be customized to suit the individual needs of a given enterprise.

4.1.3 Preparation for Implementation and Implementation Models

Literature provides numerous models of process implementation in enterprises in line with the Lean Manufacturing concept: Feld (2001), Liker (2004), Liker and Meier (2006), Oakland (2014). Indeed, one could conclude that every company seems to have its own implementation path, as confirmed by J.K. Liker (1998) in his analysis of case studies and experiences gathered in various companies from around the world.

Process implementation viewed on the example of Toyota was presented by Liker (2004, pp. 85–168) in Sect. 2—The Right Process Will Produce the Right Results. He discusses 7 of 14 management principles from the world's greatest manufactures (2 to 8), related to processes.

- Principle 2 *Create a continuous process flow to bring problems to the surface.* Work processes are redesigned to eliminate waste (muda) through the process of continuous improvement (Kaizen).
- Principle 3 *Use "pull" systems to avoid overproduction.* The pull system produces only the required material after the subsequent operation signals a need for it.

- Principle 4 *Level out the workload (heijunka).* This helps achieve the goal of minimizing waste, not overburdening people or the equipment, and not creating uneven production levels.
- Principle 5 *Stop, when there is a quality problem (jidoka).* Any employee has the authority to stop the process to signal a quality problem. Jidoka highlights the causes of problems because work stops immediately when a problem first occurs. This leads to improvements in the processes that build in quality by eliminating the root causes of defects.
- Principle 6 *Standardize tasks for continuous improvement.* By documenting the current best practice, standardized work forms the baseline for continuous improvement. As the standard is improved, the new standard becomes the baseline for further improvements, and so on. Improving standardized work is a never-ending process.
- Principle 7 *Use visual control so no problems are hidden.* Visual control methods aim to increase the efficiency and effectiveness of a process by making the steps in that process more visible. If something is clearly visible or in plain sight, it is easy to remember and keep at the forefront of the mind.
- Principle 8 *Use only reliable, thoroughly tested technology.* Technology should be pulled by manufacturing, not pushed to manufacturing.

W. Feld developed a "roadmap" to LM implementation (Feld 2001, p. 15) which divides the process into 3 phases – see Fig. 4.2.

Oakland (2014, p. 317) analyzed the correlations between the implementation Lean Management methods and the sequence thereof - Fig. 4.3.

This model can also serve as a very valuable indication as to the order in which to introduce the respective methods serving as the basis for the implementation of the process approach. As follows from our own studies, the first level should include, apart from change management and team work, also the 5S method, as its primary purpose is to maintain clean, ordered, well organized, and safe workstations, which constitute the foundation on which processes are introduced as they facilitate:

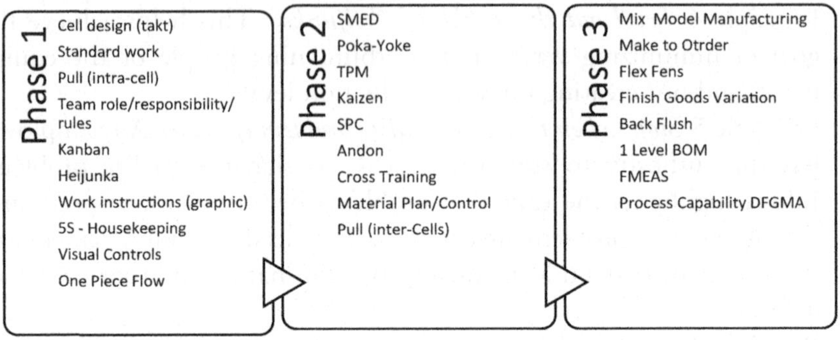

Fig. 4.2 "Roadmap" to Lean Manufacturing implementation (*Source* Based on Feld 2001, p. 15)

<table>
<tr><td colspan="6">Continuous improvement and Kaizen events</td></tr>
<tr><td>TPM</td><td colspan="2">Cellular and flow</td><td colspan="2">Pull system and Kanban</td><td>JIT</td></tr>
<tr><td colspan="2">Poka-Yoke</td><td colspan="2">Self-inspection</td><td colspan="2">Autonomation</td></tr>
<tr><td colspan="2">Batch size reduction</td><td colspan="2">Quich changeover</td><td>Layout</td><td rowspan="2">Value Stream Mapping</td></tr>
<tr><td colspan="2">Standardize work</td><td>Poin of Use Storage</td><td>Visual management</td><td>5S</td></tr>
<tr><td colspan="4">Change Management</td><td colspan="2">Teamwork</td></tr>
</table>

Fig. 4.3 Building blocks of LM (*Source* Based on Oakland 2014, p. 317)

- reduction of delays, better control of the production flow,
- higher quality, fewer errors, mistakes, and shortages,
- reduced incidence of machine breakdowns,
- better discipline and work safety

It is noteworthy that many non-production companies have also noticed the key role of the 5S method as a factor contributing to a work environment conducive to the implementation of process management (Locher 2011, pp. 120–131).

4.1.4 Factors Critical to Successful Implementation

In the case of production companies, new processes are implemented with the introduction of new products, as production processes are methods of executing technological tasks necessary to manufacture specific products. The same does not mean, however, that production companies are process oriented by default. In fact, many have relied on the functional approach to production processes for a number of years and need to be convinced to employ the processual concept. For production companies, such transition means:

- Focus on the results and management of production processes;
- Customer orientation in all departments of the enterprise by introducing the relationship of "supplier—(internal/external) customer";
- Management of production processes in terms of the determinants of competitiveness described as QCDF (quality, cost, delivery, and flexibility);
- Focus on the value stream, identification of operations contributing to value creation, possibility of optimization, reduction of nonvalue-creating operations, elimination of losses;
- Horizontal communication, reduction of the number of hierarchical tiers, a single coordinator (process owner) guarantees a clear scope of responsibility;
- Simplified processes but more comprehensive tasks of respective employees—necessitating HR development;
- Collaboration rather than competition between teams contributing to the customer value created;
- Easier quality management;
- Efficient communication and knowledge management.

Rather than focus on the implementation of the process management and its premise as such, companies opt for the introduction of comprehensive concepts such as TQM, Lean Management, and TOC, which facilitate the transition from the functional approach to process approach. The following key factors affecting application of the process approach in production companies have been identified:

- Involvement of top management, their understanding of the need for changes and acceptance of the implementation of process management as a strategic objective
- Development of an organizational culture engaging employees in problem solving and conducive to good communication and collaboration
- Monitoring the achieved results using a system of adequate individual and team-based measures
- Introduction of an incentive system targeted at rewarding employee engagement in continuous process improvement;
- Introduction of employee self-control measures and rules of personal responsibility for task execution, instead of traditional supervision and control;
- Training system to guarantee improvement of employee competences as well as familiarity with the tools for individual and team-based problem solving;
- Following good examples, dissemination of developed solutions;
- Standardization of solutions to guarantee:
    - Process stability – repeatability and reproducibility of constant result in terms of quality, productivity, costs, lead times, safety, etc.
    - Knowledge retention – when experienced employees change jobs or retire, the best working practices they developed remain in the organization in the forms of documented standard work procedures,
    - Employee training – the standard work procedures serve as the basis in training new employees and developing competences of the "old" ones,
    - Problem identification and solving – documented standard work procedures serve as the basis for the assessment of the current situation and identification of problems,
    - Continuous improvement – any process will generate wastages. But if a process is stable (performed in accordance with standard work practices), efforts aimed at improvement can begin. The current standard is also the yardstick against which the effects of improvement will be measured.

4.2 Process Implementation in Public Administration—Katarzyna Rostek

In public administration, process implementation is initiated with the transition from functional to process management. As immediately follows for the above, it is an organizationally, resource-wise, legally, and financially complex undertaking that requires a comprehensive approach accounting for a number of internal and external factors and conditions. For this reason, the structure of this chapter will deviate from that adopted in the case of production companies. Firstly, the critical determinants of implementation success will be discussed as a specific checklist to be considered before any actual operating steps are taken. The next section will focus on elements related to the organization's preparedness for the implementation and the proposed methodology contributing to the safety of the endeavor. The chapter will conclude with an overview of a number of implementation strategies to present the possible paths that can be followed, the scope and organizational potential of this type of projects.

4.2.1 Factors Critical to Successful Implementation

Based on the results of our own and secondary research, groups of factors critical to the success of a new process's implementation in public administration were identified and classified as follows:

- management – pertaining to the strategic aspect of managing and organization, including the influence of broadly understood environment (legal, economic, social, political),
- organizational – pertaining to the operational aspect of managing an organization, including the strong impact of organizational culture and the human factor,
- executive – pertaining to the scope of process implementation (organizational, technical, resource-related) and the related functional areas of the organization.

A more in-depth elaboration on the aforementioned groups would require a comprehensive analysis of the given implementation environment. Nonetheless, example studies can be cited that more specifically analyze factors critical to the success of the process approach in a specific public administration context (Syed et al. 2018, pp. 11–14):

– management factors:
  - strategic clarity and alignment – strategic vision, mission, goals, and operational plans of a public sector organization that are aligned with the political and social aims established by the government leadership,
  - external environmental – global and local political, social, and economic macro factors that influence a public sector transformation endeavor,
  - enlisting customer and stakeholder support-involvement – a systematic approach incorporating active stakeholders and customer participation to continuously monitor and identify system and service needs for process improvement,

– organizational factors:
  - project management and project management skills – a robust project management methodology and development of leadership, interpersonal, planning, financial, communication, negotiation and change management skills for staff involved in BPM projects,
  - ICT awareness – training and development initiatives to develop technical competence of public sector employees in order to capitalize on potential use of ICT systems,
  - culture – organizational core value system, shared beliefs and norms that differentiate an organization in an industry,
  - alleviation of downsizing fears – employees resistance to change from the effects of actual or expected outcomes related with BPM and IT implementation,

– executive factors:

  - choosing the BPM team – a coherent cross-functional unit skilled with relevant technical and management techniques to lead the BPM implementation in an organization,
  - empowerment – a managerial process to ensure provision of required delegation of authority, responsibility and accountability to the core BPM team,
  - ICT infrastructure – physical hardware, system architectures, software systems, technology policies and standards, and management processes required for seamless collection and dissemination of information,
  - IT- BPM governance – management practices and policies, span of management, and standard operating procedures to guide ICT operations and BPM projects in a public organization.

The above sample analysis can serve as a starting point in the development of a dedicated verification survey for a specific organization. Notably, it is important that such a survey should correspond to and elaborate on the applicable process implementation framework (Fig. 4.4) which, in turn, should distinguish the spheres of: environment (people, organization, technology), knowledge (theoretical and methodological basis), and implementation (created elements, tools allowing evaluation of the created elements).

The organizational culture constitutes a particularly significant environmental factor as numerous studies have suggested that organizations making an effort to change their current inward orientation into a more customer-oriented mentality are more likely to successfully implement the process approach (vom Brocke et al. 2011, pp. 357–378). Such a cultural transformation boils down to the establishment of standards and values consistent with the primary priority—focus on the customer and the customer value created (in the case of public sector organizations, the latter is synonymous with public value). Such a change is feasible only if the institution is ready to become an open organization, i.e. it (Snopko 2011, pp. 34–35):

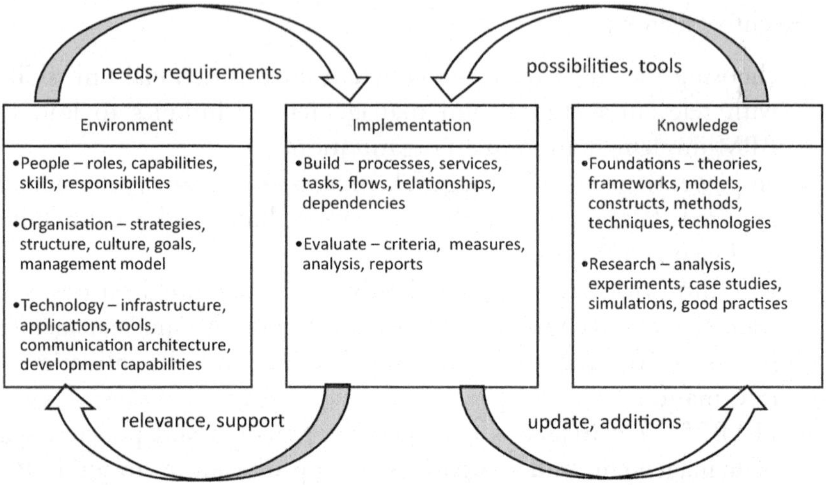

Fig. 4.4 Processes implementation framework (*Source* Based on Hevner et al., p. 80)

- is willing to individualize its activities relative to the need for development and self-fulfillment,
- shows no tendency towards unification and stabilization of its codes of conduct,
- is ready to implement justified changes,
- is not inflexibly bound to tradition (in particular preservation of the hierarchy of functional dependencies),
- knows and appreciates modern methods and approaches in management and is willing to implement the same in its own management model.

The listed characteristics of an open organizational culture are among the most important aspects in the context of implementing the process approach in public administration institutions. Of the same, the greatest significance must be attributed to the willingness to change. Particularly common and strong impact is related to legal changes that directly and fundamentally impact the operations of public sector organizations. Laws determine their activity, structure, and personnel, often also limiting their capacity for operating flexibility. In this context, it becomes vital to facilitate the individual development of employees who need to be prepared for the problems and challenges related to implementation of the process approach, but also inform them of the benefits and possibilities that the same promises.

Additionally, M. Hammer and J. Champy emphasized that processually structured organizations benefit from having employees who are not only capable of planning and executing the assigned tasks but are also capable of creating a suitable work environment for themselves (Hammer et al. 1993, p. 86). Naturally, such development is possible only if the employer provides its personnel with access to specific courses and trainings related to the process approach as well as tools supporting process management. In this context, it becomes important for an organization to have adequate solutions in place (methods, techniques, IT tools) capable of effectively supporting employees responsible for managing processes at each stage of their respective lifecycles. With reference to Fig. 4.4, one should also note the value of creating, maintaining, updating, and disseminating the organizational knowledge base with regard to subsequent project implementations or newly updated processes as a natural way of improving their quality and effectiveness.

4.2.2 Preparation for Implementation

Given the specificity of public administration, preparations necessary before implementation must entail:

- identification or reconciliation of processual and procedural standards, document templates,
- provision of personnel duly trained in terms of process management, operation, and control,
- selection of process notation and training personnel in its correct use,
- provision of a proper IT environment, including suitable tools supporting process modeling, analysis, control, and documentation.

It seems that standardization is an aspect inextricable from process implementation in public sector institutions. This is because, as demonstrated in the previous chapter, the economic and legal context is the first that has to be considered when modeling a process (Fig. 3.9). By accounting for this dimension, an organization is inevitably left with very limited or downright no room for change, which results in strict definition of a given process's execution, i.e. its standardization. In this sense, process standardization takes the form of a fixed order, with any departure therefrom simultaneously indicating greater process flexibility and lower level of standardization (Ćwiklicki et al. 2015, p. 43).

Process standardization always results from the economic and legal context, but may also touch upon the organizational perspective, particularly in entities with an extensive and complex organizational structure. A good example is provided by the Warsaw University of Technology, where the implementation and addition of new functionalities to the institution's centralized USOS system (University System for Studies Management) required prior adoption of common processual, procedural, and documentation standards by 20 university departments utilizing the system. To this end, a separate administrative unit had to be established—the President's Commission for Standardization of Studies Management and Documentation Processes—tasked with reconciling central processes and then submitting them to the University's IT Centre and deans of respective departments. The commission is composed of representatives of all departments, and the consultations conducted by this body are binding for the entire university. Hence, the entity is simultaneously tasked with developing standards for processes and procedures, as well as their further adaptation, improvement, and development.

Process standardization is an effort that needs follow a certain universal sequence of steps (Fig. 4.5), with the first thereof entailing process Identification in terms of:

- identifying organizational processes relative to the adopted classification, e.g. as strategic, primary, or supporting,
- definition process goals,
- defining the external and internal customers associated with the processes,
- assigning process ownership and corresponding scopes of authority and responsibility,

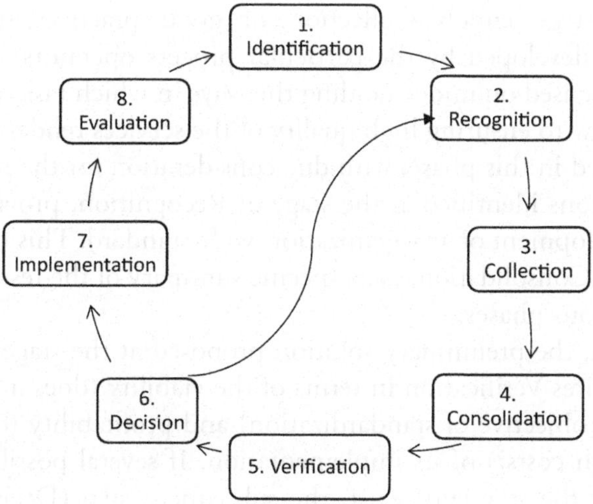

Fig. 4.5 Method of process standardization (*Source* Own elaboration)

- determining correlations and dependencies between respective processes,
- determining correlations and dependencies between process operators (organizational units, functional positions).

The subsequent step is Recognition, which focuses on the context of process execution: economic and legal, organizational, and domain-related. The objective at this stage is to fully recognize and account for all organizational requirements, social expectations, and economic/political limitations. To that end, it should involve:

- identification of pertinent economic and legal factors,
- determination of social expectations and identification of public value attributable to the processes,
- determination of organizational and domain-related factors affecting the execution of processes,
- defining adequate control measures of process correctness and efficiency.

The next stage entails Collection of good practices, models and approaches developed by the particular process operators. The stage is primarily focused on understanding the ways in which tasks are executed with the view to ensuring high quality of the services rendered. Information collected in this phase, with due consideration for the requirements and limitations identified at the stage of Recognition, provide the basis for the development of an organization-wide standard. This takes place at the stage of Consolidation, i.e. a specific summary of the results obtained in the previous phases.

Naturally, the preliminary solution proposed at the stage of Consolidation requires Verification in terms of the viability (does it really satisfy the defined objective of standardization) and profitability (do the benefits outweigh costs?) of its implementation. If several possible solutions (versions of the standard) exist, the subsequent step (Decision) entails selecting one of them based on evaluation and choice criteria approved by the management and correlated with the organization's criteria for performance assessment. It is also possible that none of the proposed variants proves satisfactory, in which case it may be necessary to revert to the first phase of analysis and collect additional information to account for any elements that may have been missed the first time around. Choices should not be rushed and the decision to supplement the findings made at earlier stages should not be perceived as a failure if the ultimately developed system is to be efficient and acceptable for all process operators within the organization.

Successful selection of the new standard allows the organization to proceed to the stage of Implementation. Specific considerations related to the implementation method will be discussed in the subsequent chapter but is it noteworthy at this point that this step also involves disseminating information and conducting trainings. It is crucial that all persons involved in the process and obliged to adhere to the new standard be informed of the implementation, fully and in a timely manner, and have all the skills necessary to effectively employ and utilize the standard. This will help to reduce the problems related to the quality of public services and resistance to change, both of which are significant obstacles when adopting a new standard in an organization.

Careful preparation of the new standard ought to ensure its correct implementation and operating effectiveness. Nonetheless, control and Evaluation of the same over time will be necessary, for instance due to new circumstances that may emerge in the economic, legal, or organizational context. For this reason, Evaluation ought to be performed in accordance with a predefined schedule (e.g. annually) as well as outside the same as needed, e.g. when legal provisions regulating execution of the given process are amended.

Consolidation of a process standard depends on its competent Implementation, which in turn runs the risk of failure. One of the ways of limiting the same is to take advantage of a verified implementation method.

4.2.3 Implementation Method

The implementation of new processes in an organization is undoubtedly a complex and highly risky endeavor. It is therefore advisable to take advantage of recommendations offered by companies specializing in the implementation of the process approach based on their practical experience and derived from the contexts of various types of entities. For instance, Rhythm Systems identified the following factors conditioning the successful implementation of a new process in an organization[1]:

- implement the process with a clear objective—starting with a statement of objectives, it is easy to understand what process is needed, how it should be made, and what result is needed to achieve. All this clearly points to the business process that needs to be updated and the business reason why it needs to be done. This establishes clear objectives to determine a common vision around the new process and its implementation,
- get process input from across the organization—most processes involve multiple organizational entities and work placements. Gaining the input from these participants on how the new process might make

[1] https://www.rhythmsystems.com/blog/how-to-implement-a-new-process-company-wide < read 2020-07-29>

things easier for them, will not only make them feel that they have been heard, but also more likely to see the value of the new process and become early adopters. It is not just the human resources, strategic management or the senior management team's responsibility—all of the cross-functional process participants should be represented in the implementation plan if the process implementation is to be successful,

- clearly communicate about the implementation process—it must be clearly communicated that a process has been improved or newly established. It is necessary to specify the value that the new process will provide to the organization and all of the participants involved. The performing steps should be clear for all participants and they should be aware of the short- (improvement context) and long-term (strategic context) benefits of introducing the change,
- drive new business process adoption—the most successful process implementations are assigned to one person—clearly defined owner—who is in charge of the key management components pertinent to the process implementation. These components need to be part of the performance management metrics so it can be clear that process implementation has the desired effects. Executive sponsorship can also be helpful in driving adoption to get the project implemented successfully,
- make adjustments to the process—if any part of the new process is not working, it is necessary to have an open dialogue and make any necessary adjustments. New processes have a chance to become acceptable standards if they truly provide the value as described in the original objective statement. It is therefore highly advisable to start implementing process changes in an ongoing cycle.

Based on the above recommendations, one can conclude that implementation of a new process ought to adopt and adapt tested implementation procedures used, for instance, in the context of software engineering, as there too issues related to clear identification of the primary objective, detailed identification of requirements with due account for the interests of all stakeholders, and ability elicit support for the proposed implementation are pivotal elements determining successes. Figure 4.6 presents

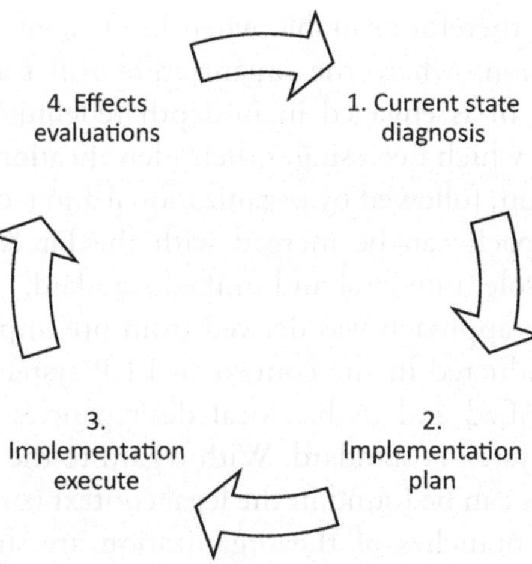

Fig. 4.6 Process implementation cycle (*Source* https://sceweb.uhcl.edu/helm/RationalUnifiedProcess/process/workflow/environm/co_imppr.htm)

a proposed scheme adopted from the Rational Software Corporation approach.

The four primary phases of implementation form a cycle (Fig. 4.6) that can, if needed, be executed iteratively, in stages. This approach allows for the introduction of evolutionary changes in organizational processes, and at the same time, leaves enough time for necessary adaptation.

4.2.4 Implementation Strategy

Any given implementation procedure can utilize a variety of methods depending on the scope and scale of the introduced changes. Three primary methodologies can be distinguished:

- Detailed specification,
- Fit & Gap,
- Agile.

The detailed specification approach entails comprehensive identification and definition of processes by following all the stages presented in Fig. 4.5—it is therefore suitable when implementing a process standard or in situations where the organization still transitioning to the process approach or is engaged in in-depth redefinition of its existing process structure, which necessitates their identification, standardization, and documentation, followed by organizational and technical implementation. The approach can be merged with the Fit & Gap method or employed as the sole, universal and unified standard.

The Fit & Gap approach was derived from pre-implementation analyses typically conducted in the context of ERP systems. It involves the definition of justified and usable local discrepancies in the process of implementing a systemic standard. With regard to the process approach, such discrepancies can be found in the legal context (so-called legal gaps), where respective branches of the organization are subject to different local laws (or different interpretations thereof), and the domain context (local gaps), where local organizational units define their own, unique processes (e.g. due to providing a unique public service specific to the given location). All the observed differences (gaps) are identified but the objective in this approach is also to decide which of the same, and to what extent, warrant implementation and whether such implementation is to be local or global, the latter leading to the adopted gap becoming an organizational standard (fit). This approach is a method of reconciling/negotiating process standards, which contributes to a better intra-institutional transparency and involvement of all process stakeholders in its amendment and improvement. Besides, the negotiatory approach promises to not only more thoroughly prepare an organization for the acceptance of change, but also limit later resistance to change from employees who would otherwise have no say in the identification/selection of the final process formula.

In the case of administrative processes, one major obstacle to implementation is the fact that only some can be classified as routine. At the same time, one can encounter cases requiring individualized processing, negotiation, as well some that are simply poorly structured. Each of the above are characterized by difficult standardization, and consequently also automation. In this case, the solution may be to adopt the Agile approach to process implementation (Hinkelmann et al. 2006; Feldkamp et al. 2007), whose main premise is to ensure flexibility and ability to cope with changes and exceptions that cannot be handled with traditional methods. However, this approach requires high efficiency and competence from all members of the implementation team tasked with adopting and then individually adapting a given standard to the requirements of the situation at hand. If extended to encompass the entire cycle of process management, the approach can combine processual and organizational knowledge with familiarity with the socio-legal context wherein the organization functions, yielding activity consistent with the developed business principles and process models that are flexibly adaptable but within closely controlled limits. All in all, this leads to more flexible and agile business processes by integrating business process management with business rules in three different ways (Feldkamp et al. 2007, p. 37):

- variable activities are linked to action-enabling rules. This leads to flexible process execution, because the final process flow is determined at run-time. In contrast to adhoc workflows, where a human participant decides to deviate from a predefined process flow, traceability and compliance can be controlled by a set of rules,
- resource allocation is more flexible. For example, a participant can be selected based on individual skills instead of general role specifications,
- intelligent branching is supported by business rules representing expert knowledge.

In addition, violations of constraints and guidelines are determined immediately at run-time, avoiding expensive repair actions.

It seems that the Agile approach may indeed provide a remedy for non-routine processes in public administration as its goal is (Hinkelmann et al. 2006, p. 33):

- to soften the strictness of business process structures to be handled: exceptional and unpredictable situations, high variability of input and output, highly complex tasks, innovative, long-running processes, new experiences,

- to formalize the weakness of work processes knowledge to be increased: decision support (providing comprehensive but task-oriented knowledge), compliance with rules and regulations, consistency of decisions, transparency and traceability of work, process automation.

The eGovernance platform rendering public services to Swiss citizens was developed and deployed in accordance with the Agile approach (Hinkelmann et al. 2006, pp. 33–41). The project, which also entailed implementation of the process approach in Swiss public administration institutions, managed to accomplish three goals: (1) overcoming barriers stemming from the federal structure of administrative division and the resulting differences in terms of the social and legal situation (2) modeling and partially automating poorly structured but task-oriented processes, and (3) modeling and partially automating processes requiring expert knowledge. The breadth of those accomplishments corroborates the potential of the Agile approach in implementing processes that could not previously be properly executed due to their high requirements in terms of flexibility and variability.

4.3 Comparison and Recommendations—Anna Kosieradzka and Katarzyna Rostek

While process modeling can be performed with the use of the same tools and notations in various types of organizations, the implementation of processes in each of the discussed groups varies significantly, as clearly evidenced in the aforementioned juxtaposition of the focal points characterizing this stage.

4.3.1 Implementation Strategy

Production companies focus primarily on reorganizing, redesigning, and optimizing processes in terms of their quality and efficiency as well as in light of unavoidable changes to the production process. Hence, their activities are aimed towards approaches that facilitate the accomplishment of the above, such as TQM, LM, or TOC. It should be noted that production companies stand as a model for other types of organizations, paving the way in areas of continuous improvement and development of the process approach with a view to maximizing performance and productivity.

Public administration is an area where the process approach is only currently being introduced. Therefore, in this case process implementation tends to be an entirely new project radically altering the management model and functional operation of the organization. In many cases, this necessitates in-depth reorganization in terms of the organizational culture, decision-making and functional hierarchy, and even the structure and operating policy of an institution. Given the above, when discussing implementation strategy in the context of the public sector, one has to consider a highly complex process whose scope suggests the applicability of a detailed, supplementing, or agile project strategy. The ultimate goal is to ensure high efficiency of the entire project in order to allow the organization to operate in a project-based manner rather than rely solely on its functional structures.

4.3.2 Implementation Methods

The methods of implementing processes in production companies provide an elaboration on the adopted concept (e.g. TQM, LM, TOC) as well as ensure its desired quality and efficiency. The methods focus on primary and supporting processes as the same are the object of the concepts in question. This approach yields the expected results and facilitates the achievement of the adopted goals (strategy and optimization). The entire organization is guided towards activity conducive to its primary processes with the additional guarantee of optimizing resource consumption, eliminating mistakes and defects in production. As in the case of the strategic approach, also detailed methods facilitating the achievement of strategic goals may be treated as specific models that can be adapted by developing and improving organizations of other types.

When implementing the process approach from the ground up, public administration institutions opt for well-known and proven methods of project management. It is important that the utilized methods are adequate to the adopted strategy and show verified reliability in the context of complex reorganization projects. A good source of such solutions and their adaptations is provided by models and methods employed in complicated IT projects, which is why this chapter discussed one of the same.

4.3.3 Preparation for Implementation and Implementation Models

In the case of production organizations, the preparatory stage of implementation entails, above all, identification, recognition, and adaptation of implementation models associated with the selected production management concept, with a view to customizing the approach to match the situation of the given organization, on the one hand by adapting the same to the internal and external situation thereof, and on the other, by maximizing its benefits.

In the case of public administration, this stage aims to ensure the organization's preparedness for the introduction of processes and process management. This involves conducting a very thorough analysis of the organizational structure and its functional capacity, including in terms of identifying the scope of differences between the respective organizational units relative to operational performance and decision-making, and standardizing relevant processes and procedures as required. This is particularly important if the implementation of processes and associated management techniques is to also facilitate standardization and normalization of functions and tasks as well as to guarantee organization-wide comparability with regard to quality and effectiveness of customer services.

4.3.4 Factors Critical to Successful Implementation

As regards the key success factors applicable to implementation, they remain fairly similar regardless of the type of organization as they are determined by the very nature of the process approach. However, they do differ in terms of their critical significance. In the case of production companies, such factors play a secondary role as it goes without saying that the operations of such organizations are, by principle, based on processes, with the main focus invariably on their optimization and improved organization, rather than their emergence as such. Meanwhile, in the case of public administration such factors should be treated as a specific checklist that needs to be verified before making any decisions regarding implementation. The results of such an assessment will determine the scope of efforts necessary to prepare the organization for implementation, and only once all such requirements are met can one consider the choice of specific implementation methodologies and strategies.

References

Cox, J. F., & Schleier, J. G. (Eds.). (2010). *Theory of Constraintr Handbook*. New York: McGraw-Hill.

Ćwiklicki, M., & Walczak, M. (2015). Modelowanie Procesów Jako Podstawa Standaryzacji Kosztów w Jednostkach Samorządu Terytorialnego. *Nauki o Zarządzaniu, 3*(24), 41–57.

Feld, W. M. (2001). *Lean Manufacturing. Tools, Techniques, and How To Use Them*. Boca Raton: The St. Lucie Press.

Feldkamp, D., Hinkelmann, K., & Thönssen, B. (2007, October). Kiss–Knowledge-Intensive Service Support: An Approach for Agile Process Management. In *International Workshop on Rules and Rule Markup Languages for the Semantic Web* (pp. 25–38). Berlin, Heidelberg: Springer.

Hammer, M., & Champy, J. (1993). *Reengineering the Corporation. A Manifesto for Business Revolution*. New York: HarperCollins Publishers.

Hinkelmann, K., Probst, F., & Thönssen, B. (2006). Agile Process Management: Framework and Methodology. In *AAAI Spring Symposium: Semantic Web Meets eGovernment* (pp. 33–41).

Jones, D., & Womack, J. (2011). *Seeing the Whole Value Stream*. Brookline: Lean Enterprise Institute.

Liker, J. K. (2004). *The Toyota Way. 14 Management Principles from the World's Greatest Manufacturer*. New York: McGraw-Hill.

Liker, J. K. (Ed.). (1998). *Becoming Lean*. Portland: Productivity Inc.
Liker, J. K., & Meier, D. (2006). *Toyota Way Fieldbook*. New York: McGraw-Hill.
Locher, D. (2011). *Lean Office and Service Simplified*. New York: CRC Press.
Oakland, J. S. (2014). *Total Quality Management and Operational Excellence*. New York: Routledge.
Rother, M., & Harris, R. (2001). *Creating Continuous Flow*. Brookline: Lean Enterprise Institute.
Rother, M., & Shoock, J. (2009). *Learning to See*. Lean Enterprise Institute, Brookline: Value Stream Mapping to Create Value and Eliminate Muda.
Snopko, J. (2011). Kulturowe Uwarunkowania Dokonywania Zmian Organizacyjnych w Urzędach gmin. *Zarządzanie Publiczne, 14*(2), 27–39.
Syed, R., Bandara, W., French, E., & Stewart, G. (2018). Getting it Right! Critical Success Factors of BPM in the Public Sector: A Systematic Literature Review. *Australasian Journal of Information Systems, 22,* 1–39.
Vom Brocke, J., & Sinnl, T. (2011). Culture in Business Process Management: A Literature Review. *Business Process Management Journal, 17*(2), 357–378. https://doi.org/10.1108/14637151111122383.
Womack, J. P., & Jones, D. T. (1996). *Lean Thinking*. New York: Banish Waste and Create Wealth in your Corporation. Simon & Schuster.

5
Process Verification

Dr. Anna Kosieradzka and Katarzyna Rostek

5.1 Process Verification in Production Companies—Anna Kosieradzka

The below section will discuss the theoretical and practical aspects of process evaluation and verification, presented on selected examples of production organizations derived from both our own and secondary research results.

5.1.1 Characteristics of Process Measures

Efficient and effective management of production processes needs both managers and other staff members to have access to reliable financial and non-financial measures related to the processes they supervise and perform (Kaplan and Cooper 1998, pp. 81–82). The non-financial measures most commonly employed in the context of production processes include: effectiveness, efficient use of production resources,

A. Kosieradzka and K. Rostek, *Process Management and Organizational Process Maturity*,
https://doi.org/10.1007/978-3-030-66800-6_5

production quality, productivity, production cycle, supply cycle, and stock level, while the main financial measure is profitability.

The measures of production efficiency and effectiveness are closely correlated, as illustrated in Fig. 5.1 below.

Effectiveness is measured by the degree to which a given process does what it is intended to do. The measure focuses on the output of the production process.

Quality is the degree to which the requirements and expectations of customers are met by the system. The primary question asked in the context of quality is: *Is the customer satisfied with the product (service)?* It therefore contains the effectiveness aspect.

Efficiency is the degree of resource consumption. It pertains to production process inputs and measures the efficient use of available resources.

Productivity is the ratio between the number of products generated in the production process and the amount of input resources consumed in the same. The numerator of the expression contains the aspects of effect and quality: products meeting quantitative, qualitative, and time-related requirements of the customer. The denominator contains the aspect of efficiency: resources consumed and/or used.

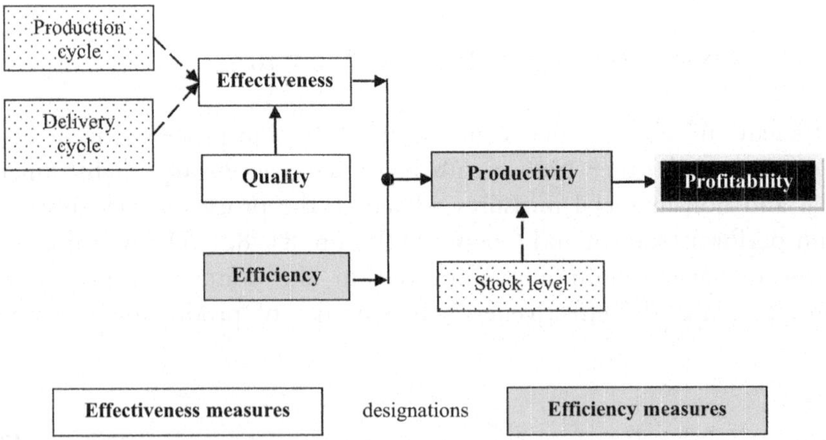

Fig. 5.1 Correlations between process effectiveness and efficiency measures (*Source* Own elaboration)

The operational measures strictly related to production processes include: production cycle, delivery cycle, and stock level. Their respective values are affected by solutions adopted in the spheres of organization and management of production processes. The production cycle and supply cycle directly and indirectly influence the effectiveness of production processes, while the stock level impact productiveness.

Profitability is the ratio between the profits generated by the production system and the sales figures, assets allocated, or equity. Typical financial profitability measures include: net profit, return on assets, and return on equity.

5.1.1.1 Measures of Production Process Effectiveness

Effectiveness of a production process is measured by the degree to which the process achieves its intended results. At least two aspects need to be considered if one is to viably measure effectiveness: quantity and timeliness, i.e. *Was the product designed, manufactured, and delivered in the required quantity and within the required deadline?*

Effectiveness is closely associated with planning. The measure of effectiveness is the ratio of planned to actual performance.

$$Effectiveness = \frac{production\ performed\ [unit]}{production\ planned\ [unit]} \bullet 100\% \quad (5.1)$$

Timeliness of order completion can also constitute a measure of production effectiveness:

$$Timeliness = \frac{number\ of\ orders\ completed\ on\ time}{total\ number\ of\ orders\ completed} \bullet 100\% \quad (5.2)$$

Effectiveness is closely related to organization of the production process, i.e. the value stream which—as already mentioned—is a sequence of activities aimed at providing customers with products consistent with their expectations. The value stream includes in particular (see the MTO model—Make-to-Order): the complete design process—from the product concept to implementation of production, and the production process—from procurement of raw materials and materials to delivery of the finished product. In repeatable production (see MTS model—Make-to-Stock), the value stream is limited to the production process itself. To distinguish the two situations, in the case of the MTO model the concept of delivery cycle (lead time) was introduced to indicate the period from the customer's order to the eventual delivery of the ordered product, whereas in the case of MTS, the delivery cycle fully overlaps with the production cycle.

Production cycle (duration of the production process) is the period of time from the start of production of a given output to the completion thereof. It is the period during which the input material is processed in the respective production operations and shaped into the finished product. This duration can range anywhere between several seconds—e.g. simple, mass-produced goods, and several years—manufacture of individual, highly complex products e.g. ship or building.

Two main periods can be distinguished in the context of a production cycle: work and breaks. The work period covers: the technological period of the production cycle—i.e. the duration of technological operations, and other activities: control, transport, and maintenance. The period of breaks includes breaks imposed by the calendar, organization of each working day, as well as the adopted production process, e.g. forms of production organization, batch sizes, parallel partial processes, synchronization of work between respective workstations, etc.

The production cycle is analyzed in the context of the value stream. In particular, it entails identification of three types of activities:

- Value-adding activities–all operations that result in developing properties and functionalities of the product in accordance with the customer's requirements (primarily technological operations),
- Muda Type 1–non-value-adding activities in the processes that are nonetheless necessary for the end customer, for example, inspection and safety testing.
- Muda Type 2–non-value adding activities that are unnecessary for the customer, for example wait time, production of defective products.

Optimization of the production process ought to be primarily targeted at the identification and elimination of type 2 Muda. The effects of such efforts can be evaluated by calculating the W_t index expressing the ratio of the technological period duration in the production cycle (O_t) to the production cycle as a whole (C):

$$W_t = \frac{O_t}{C} \bullet 100\% \tag{5.3}$$

This should be followed by identification and reduction of type 1 waste, i.e. nonvalue-adding operations necessary to facilitate production (e.g. transport, control, handling). The effects of these efforts can be measured with the W_y index: ratio of value-adding period duration (O_W) to the production cycle as a whole (C).

$$W_w = \frac{O_w}{C} \bullet 100\% \tag{5.4}$$

The aim is to maximize the values of both those indices.

Quality is difficult to define as it relates to the customer's subjective evaluation of the degree to which a given product meets his expectations. According to PN-EN 9000:2006

"quality is the degree to which the collective inherent properties of a product meet the requirements of the customer".

Management of quality assurance processes requires a system of continuous measurement, observation of various parameters, data collection, and quality evaluation presentation. Most quality measures are focused on determining the satisfaction level of external customers, whose increase does not, however, have to stem directly from the quality of the company's internal processes.

The most commonly used measures of process quality include:

- rejection index, relating the number of products rejected in the process due to quality defects (B) to the overall number of products manufactured (P):

$$W_b = \frac{B}{P} \tag{5.5}$$

This is often expressed relative to a million opportunities to make an error (produce a defective output), in which it is referred to as DPMO *(defects per million opportunities)*. Its values can be placed on the sigma quality scale used in under the Six Sigma concept (Pande et al. 2000, p. 28)

- index of "good" products (i.e. products consistent with customer requirements)—it complements the indications of the rejection index and expresses the ratio of good products (D) to the overall number of products manufactured (P):

$$W_d = \frac{D}{P} \tag{5.6}$$

This index usually utilizes the "good at first go" approach, where good products are only those that do not require any corrections or modifications.

- returns index, i.e. the ratio of products returned by customers as defective (R) to the overall number of products manufactured (P):

$$W_r = \frac{R}{P} \tag{5.7}$$

5.1.1.2 Measures of Production Process Efficiency

The measures of production process efficiency can be further divided into efficiency and productivity.

Efficiency is measured with regard to the consumption of allocated resources (employees, fixed and current assets, energy, materials, floor space, etc.), i.e. the ratio of resources used to total available resources. Examples of such efficiency measures include:

- machine worktime use

$$W_m = \frac{used\ (productive)\ machine\ worktime\ [h]}{available\ machine\ worktime\ [h]} \bullet 100\% \quad (5.8)$$

- employee worktime use

$$W_p = \frac{used\ (productive)\ employee\ worktime\ [h]}{available\ employee\ worktime\ [h]} \bullet 100\% \quad (5.9)$$

- material yield

$$W_{mat} = \frac{finished\ product\ weight\ [kg]}{weight\ of\ material\ used\ in\ production\ [kg]} \bullet 100\% \quad (5.10)$$

Productivity is understood as the relationship between the outputs (goods and services) generated by a production system and the inputs (labor, tangible and financial assets, materials, energy, information) necessary to create said outputs (Kosieradzka 2012, pp. 20–23). The general Concept of productivity is presented in Fig. 5.2, where:

Q^I_{ri} – quantity of type r input consumed in the period i,

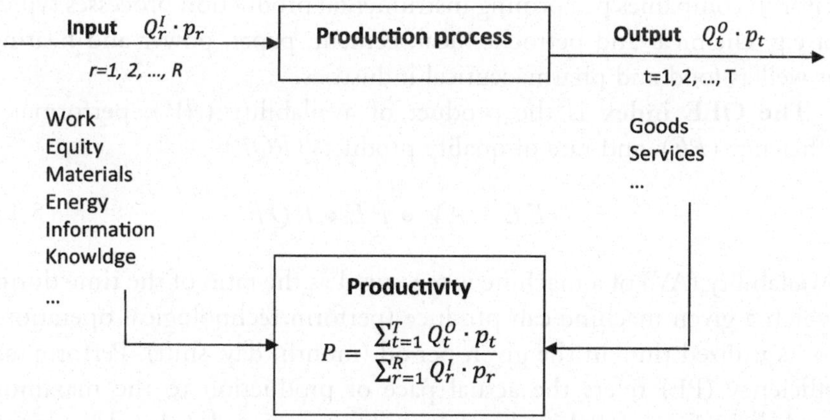

Fig. 5.2 The general concept of productivity (*Source* Kosieradzka 2012, p. 20)

Q^O_{ti} – quantity of type t output produced and sold in the period i,
p^O_{ti} – the unit cost for each input type,
p^I_{ri} – the unit price for each output type
T – 1 – number of different types of output
R – number of different types of input,

Productivity is the most important measure for production processes as it relates their overall output to resources consumed in the same.

As follows from the above definition, there are hardly any limits as to the number of productivity measures that can be adopted as any index constitution any ratio of a production output (number of products manufactured and delivered to customers, value of production sold, added value generated in the company, etc.) to the quantity of resources consumed or used to achieve that output (e.g. number of employees, cost of labor, worktime, value of fixed and current assets allocated to the activity, quantity and value f materials and energy consumed, etc.)—will constitute a productivity measure.

One specific type of productivity measure applicable to the equipment (machines and devices) utilized in the process is the so-called OEE index (Overall Equipment Effectiveness) – (Nakajima 1988, pp. 21–29). OEE constitutes a commonly used, comprehensive and synthetic indication in the evaluation of the operating effectiveness of machines and their ability to meet quality requirements. It serves as a primary productivity indicator in companies performing instrumental production processes typical of e.g. chemical and petrochemical, cement, paper, power and heating, as well as food and pharmaceutical industries.

The OEE index is the product of availability (*AV*), performance efficiency (*PE*), and rate of quality products (*RQP*):

$$OEE = AV \bullet PE \bullet RQP \tag{5.11}$$

Availability (AV) of a machine is measured as the ratio of the time during which a given machine can produce (perform technological operations) to its utilized time in the given period (month, day, shift). Performance efficiency (PE) refers the actual pace of production to the maximum level thereof generated in the absence of any micro-breaks, slow-downs, and interruptions. The rate of quality products (RQP) is the ratio of the quantity of products meeting quality requirements to the overall number of products manufactured. It indicates the worktime utilized for defect-free production against time wasted due to defects.

The advantage of OEE stems from the fact that this its value is expresses as percentage, which allows competitive efficiency analyses for equipment (individual machines, production lines) utilized in processes conducted in different companies, even countries.

AT this point, one should also consider the correlation between productivity and profitability as both of these concepts express a ratio or production system inputs to its outputs (Fig. 5.3). Firstly, it should be observed that productivity is a measure applicable to processes, whereas profitability to the enterprise as a whole. Productivity measures the efficiency of production processes, and as such is best suited to short-term analyses.

Business profitability measured relative to the return on sales (commercial profitability) and return on assets (economic profitability)

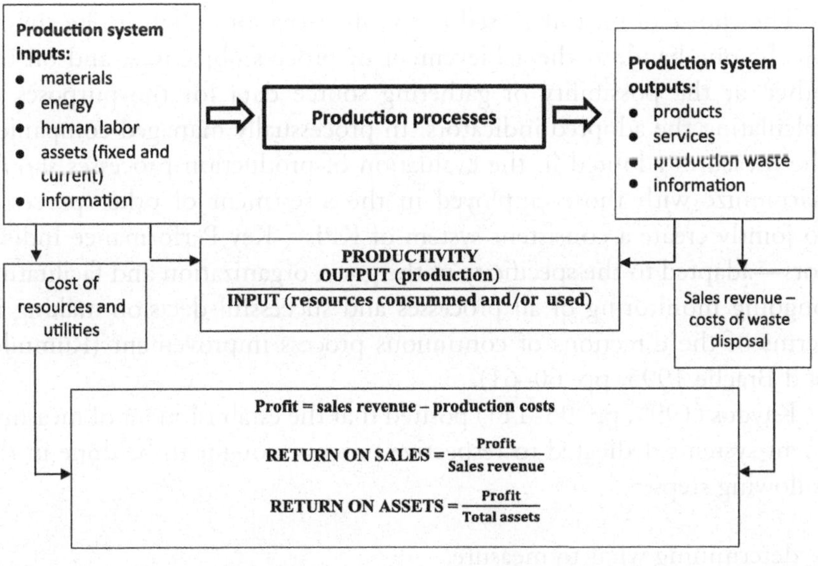

Fig. 5.3 Correlation between process productivity and business profitability (*Source* Own elaboration)

is a factor conditioning the survival and development of an enterprise, significant in the long-term perspective. Both productivity and profitability indices can be improved by increasing sales and reducing production costs. However, while productivity measures are directly dependent on production processes and their efficiency, profitability measures reflect the structure of profit generated from the respective areas of the company's operations: profit form core activity, financial activity, other sales, and extraordinary profit and loss balance. Given the above, profitability measures should not be treated as reflecting solely the condition of production processes.

5.1.2 Method of Defining Process Measures

As observed by (Harrington 1991, p. 82): *Measurement is the first step that leads to control and eventually to improvement. If you cannot measure something, you cannot understand it. If you cannot understand it, you cannot control it. If you cannot control it, you cannot improve it.*

The choice of measures used in specific situations ought to be aimed, on the one hand, at the achievement of process objectives, and on the other, at the possibility of gathering source data for the purposes of calculating the adopted indicators. In processually managed companies, the measures adopted in the evaluation of production processes should harmonize with those employed in the assessment of other processes to jointly create a consistent system of KPIs—Key Performance Indicators—adapted to the specificity of the given organization and facilitating ongoing monitoring of all processes and successful decision-making in terms of the directions of continuous process improvement (Rummler and Brache 1995, pp. 60–61).

Kaydos (1999, pp. 95–111) posited that the establishment of measurement systems dedicated to respective processes ought to be done in the following steps:

- determining what to measure,
- designing the data collection system,
- developing the data processing system,

- developing the system,
- validation,
- refinement.

He also drew attention to such aspects as legibility and clarity of the designed reports, and security of confidential information.

Based on our own studies, the procedure of adopting process indicators should consider answers to the following questions (Fig. 5.4):

It should be emphasized that the system of process measures ought to be developed individually for each respective enterprise, with due consideration for the applicable specifics and needs. The significant impact of the adopted measurement system on employees and their behavior was observed and aptly summarized by Goldratt: *Tell me, how you will measure me, and I will tell you how I will behave. If you measure me in an illogical way … do not complain about illogical behavior. (Goldratt* 1990, *p. 26)*

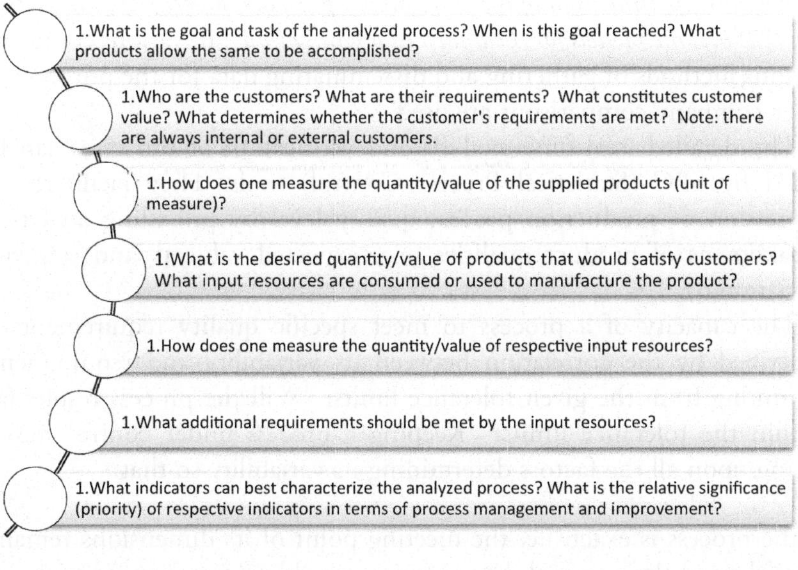

Fig. 5.4 Procedure of generating process indicators (*Source* Own elaboration)

5.1.3 Tools and Methods Facilitating Process Verification

The values of indicators as such will tell one little about the analyzed processes. They first need to be interpreted using comparative techniques, mainly in terms of the following relationships:

- comparison to accepted model values (e.g. strategic goals adopted for the respective processes);
- comparison in time (to analyze trends for changes observed for respective indicators in subsequent periods),
- comparison to other companies (e.g. key competitors, sectoral mean values).

Result benchmarking and Inter-Firm Comparison (IFC) methods allow one to evaluate the company against its competitors and identify the best processes. H.J. Harrington (1991, p. 218) defined benchmarking as "*systematic identification of the best systems, processes, procedures, and activities that become models for the analyzed organization*". In various countries, various methods of gathering and disseminating data for the purposes of inter-company comparisons are used.

The detailed, real-time analysis and verification of processes can be aided by SPC (Statistical Process Control) tools. They facilitate the expression of production process quality level by providing analytical breakdowns of product qualities (e.g. expected value μ and standard deviation σ).

The capacity of a process to meet specific quality requirements is described by the correlation between its variability and requirements stemming from the given tolerance limits: "Will the process results fall within the tolerance limits?" Keeping a process under control means acting upon all the factors determining its variability so that:

- the process is exact, i.e. the meeting point of its dimensions remains as close to the nominal dimension as possible. An exact process is one for which the mean value is equal to the nominal or target value.

- the process is precise, i.e. the dispersion of values does not exceed the acceptable tolerance limits. A precise process is one for which a relatively small distribution of values is observed relative to the tolerance range for the analyzed property (value).

Figure 5.5 illustrates the above concepts as well as the idea of meeting quality requirements defined relative to the nominal value and acceptable deviations therefrom, expressed in the form of top and bottom tolerance level (Tg and Td).

The process capability index is calculated as:

$$C_p = \frac{Tg - Td}{6\sigma} \tag{5.12}$$

The centered capability index is expressed as:

$$C_{pk} = min\left\{\frac{Tg - \overline{Y}}{3\sigma}; \frac{\overline{Y - Td}}{3\sigma}\right\} \tag{5.13}$$

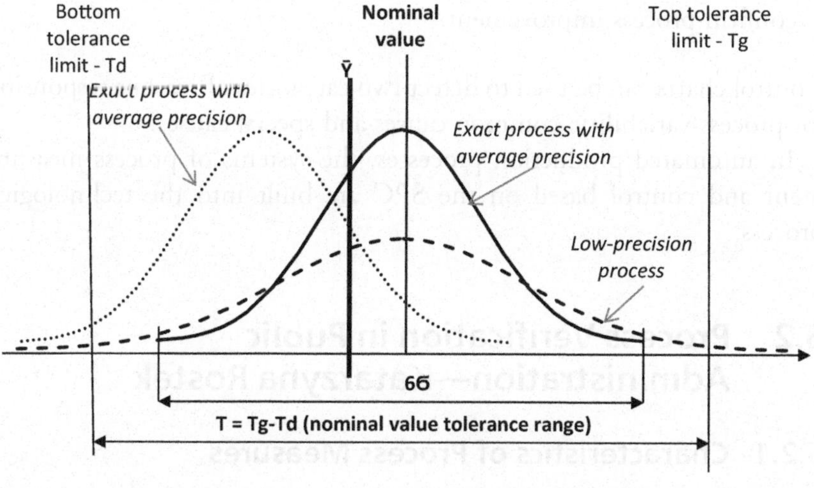

Fig. 5.5 Idea for determining process capability (*Source* Own elaboration)

The minimum acceptable value of $C_p = 1.0$ corresponds to the acceptable incidence of product defects of 0.27%. Already in the 1980s, the standard value prevalent in Japan was $C_p = 1.33$. Nowadays, the required minimum value of the capacity index tends to be industry specific. For instance, in the automotive industry $C_p = 1.67$ has been adopted (defect incidence at the level of 0.0063%). In projects implemented in accordance with the Six Sigma methodology, $C_p > 2.0$ and $C_{pk} > 1.5$ are the recommended values.

Schewart control charts are a popular tool supporting the improvement of production processes' capacity to meet quality requirements. A control chart is a unified method of recording information about process flow. It allows monitoring and analysis of the process to (Dietrich and Schulze 1999; Oakland and Oakland 2019):

- diagnose (analyze and evaluate) the short- and long-term capacity of the process,
- steer the process:
  - determine when the process is out of control,
  - determine when the process needs adjustment,
- confirm process improvement.

Control charts can be used to detect two categories of reasons responsible for process variability: common causes and special causes.

In automated production processes, the systems of process measurement and control based on the SPC are built into the technological process.

5.2 Process Verification in Public Administration—Katarzyna Rostek

5.2.1 Characteristics of Process Measures

The usefulness of measurements discussed at the beginning of this chapter is strictly conditional upon the viability of the adopted process measures. One should therefore consider the general guidelines that, although developed in the context of KPI management, can also be successfully applied in the case of the public sector:

- measures should focus on issues that are important for the decision-makers. i.e. be directly related to process goals, value generated in the process, or resources consumed in its execution,
- measures should be adapted to the service rendered or value created in the process; they should be formulated in way that is simple and understandable for the users,
- the number of measures used should not be excessive, as their usefulness decreases with greater numbers due to limitations in terms of their interpretability and parallel applicability,
- the number of measures should reflect the significance of a given process in the organization—key processes should have more measures associated with them than less important ones,
- every measure must have a specific model (standard) that will unambiguously identify the method (e.g. formula or algorithm) of calculating its value and conditions of the same (e.g. period of time subject to measurement),
- models ought to be periodically updated to ensure continued satisfaction of the decision-making usefulness condition while at the same time providing a good starting point for process improvement,
- measures should be defined in such a way that their results can be realistically (practically) used for the purposes of process verification, evaluation, and improvement—it is therefore necessary to strictly specify for whose benefit, to what extent, and for what purpose the respective measures are used,
- the costs of gathering and processing data used for calculating the measures must not exceed the benefits obtainable by applying said measures to the process—expenditures related to process verification must be rationalized to ensure adequate balance between improvement efforts and everyday operations of the organization.

Based on the above, one can specify the basic scope of a measure's definition which includes (Molenda 2016, p. 251): the name of measure, its goal (purpose), frequency of measurement, method of measurement and calculation (estimation), sources of input data, recipient of values (measurement results), person performing (responsible for) the measurements, as well as the mode and format of results and their presentation. The key factor here is the purpose that the measure is to serve as it will determine the value subject to measurement and control. The same may be any element of the infrastructure of process, including: input resources (material, information, financial) associated with any activity in the process and its outputs (products, services, information), functional elements of the process (such as efficiency or correct performance of respective process activities), as well as the executors of the respective process activities (worktime input, productivity, correctness). The method of measurement and control, its frequency and scope will consequently determine the usefulness of the given measure and viability of its application in the context of verifying and improving a process.

5.2.2 Method of Defining Process Measures

The above once again indicates the need for a systematic approach to ensure that the developed measures are feasible, rational, and useful to the organization (Fig. 5.6).

Processes subject to monitoring and control must be carefully identified and modeled (1.) to ensure that all the significant particulars are known: purpose, expected end result, process operators with their respective scopes of rights and responsibilities, scope of activities and the assigned necessary resources, as well as data, information, and results exchanged between respective activities and/or processes. Only in-depth knowledge of the process in terms of the existing bottlenecks, tangency points, information or product exchange with other processes, and elements critical from the perspective of process performance, will allow one to viably define measures consistent with the above requirements.

The next stage involves process analysis (2.) in terms bottlenecks, place, time, and extent of resource consumption, the applicable limitations and requirements, correlations and codependences with other

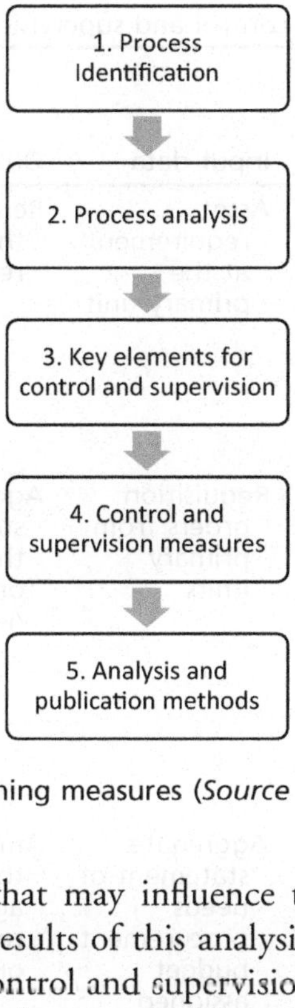

Fig. 5.6 Method of establishing measures (*Source* Own elaboration)

processes, and elements that may influence the effectiveness and efficiency of processes. The results of this analysis will help to identify the key elements of process control and supervision.

One should note that the scope of the identified key elements for control and supervision (3.) will be determined by the orientation and expectations of decisionmakers, as well as their goals and needs expressed through the decisions made. Consequently, this stage in the procedure requires particularly frequent analyses in terms of adequacy and sufficiency. Table 5.1 presents a sample, fragmentary process analysis, where each activity in the process is assigned with specific input resources (input data), products (output data), and key elements for control and supervision. Naturally, the indication was done at a specific time and by a specific person managing the process. At another time and for a different decisionmaker, the elements could have been identified differently. Regardless of the situation, however, it is important that the decisionmakers retain adequate influence on the scope and content of

Table 5.1 Key elements for control and supervision

Symbol of process activity	Name of process activity	Input data	Output data	Key elements for control and supervision
C1	Prepare a requisition order	Asset requirement at the primary unit	Requirement in the form of a requisition	Prognose needs in terms of a movable asset and compare the same to the asset level held.
C2	Collect and analyze requisition orders	Requisition orders from primary units	Aggregate statement of the organization's needs	Requirement in the form of a requisition, specifying the type and extent of staffing and unit service profiles.
C3	Plan financial resources	Aggregate statement of needs, procurement budget assigned	Amendment to the aggregate requisition order	Comparative financial analysis. Aggregate requisition order taking budget limitations into account.
C4	Submit the requisition order and inform the Director of the Police Main Logistics Office	Requisition order from a primary unit	Application for independent award of a public procurement order.	–

Source Own elaboration

elements subject to measurement and control, as it is their tasks that are to be effectively supported by the analyses of the thus obtained results.

The key elements for supervision and control ought to unambiguously specify the measures (4.) that will allow key elements of the process to be monitored in a way and within the scope expected by the decision-makers. Notably, such measures can be provided as a calculation formula (Table 5.2), procedural algorithm (Table 5.3), or a criterion with defined limit values (Table 5.4).

The last stage of the procedure (5. Analysis and publication methods) is also very important as it determines, on the one hand, the conditions and methods of establishing measured values, and on the other, the method of their publication in accordance with an approved format

Table 5.2 Measure as a calculation formula

Number and name of activity:	P1-C1. Preparation of a requisition order
Description of activity:	Preparing the requisition order for the required equipment and assets
Element requiring control and supervision:	Prognosing movable asset needs and comparing the same to the assets currently held.
Assigned indicator:	Asset requirement at a primary unit level for any given element of movable assets should be compared to the requirement prognosis calculated from *Brown's single-exponential model*: $\overline{y}_t = \alpha y_t + (1 - \alpha)\overline{y}_{t-1}$ $\hat{y}_{t+T} = \overline{y}_t$ where: $\overline{y}_t, \overline{y}_{t-1}$ – mean values calculated, respectively, for periods t (just finished) and t – 1 (preceding), $\overline{y}_t$ – last iteration of the prognosed variable α – exponential smoothing parameter – values within the range of (0,1), $\hat{y}_{t+T}$ – prognosed value of the y variable in the period *t* + *T*.

Source Own elaboration

Table 5.3 *Measure as a procedural algorithm*

Number and name of activity:	P1-C3. Planning financial resources
Description of activity:	Determining the level of financial resources necessary to procure specific assortment of assets.
Element requiring supervision:	Comparative financial analysis. Aggregate requisition order taking budget limitations into account.
Assigned indicator:	Based on the analysis in point P1-C1 and the procurement budget awarded, the aggregate requisition order is amended. If the available budget is limited, the optimum combination of assets ought to be determined. The possible combinations are presented in the *budget line* theory which distinguishes attainable and unattainable combinations under a given budget. The budget equation is: $DP = Qx * Px + Qy * Py$ where: Qx and Qy – are quantities of assets x and y, Px and Py – are prices of assets x and y. Should we assume that the total possible income is allocated to a single asset, X or Y, points along the axes determine the inclination of the budget line. With higher resources, the budget line will move up, and with lower down. The budget line may also change its inclination, depending on the purchase price of the asset. If the requirement exceeds financial capacity, needs are ranked based on priority and order timeframes, relative to the given financial capacity.

Source Own elaboration

and formula adapted to the users' needs. The stage of analysis and publication can entail the use of a variety of analytical tools. The simplest solution is, of course, to use MS Excel spreadsheets, a popular and fairly potent tool. However, far more advanced solutions are also available, for instance Business Intelligence (BI) systems.

Table 5.4 Measure as a criterion with defined limit values

Number and name of activity:	P1-D5. Does the purchase price exceed EUR 14,000?
Description of activity:	If the purchase price exceeds the equivalent of EUR 14,000, a tender for the purchase of movable assets must be announced.
Element requiring supervision:	Order amount (above or below EUR 14,000).
Assigned indicator:	Determination of the order amount: • above EUR 14,000. • up to EUR 14,000.

Source Own work

5.2.3 Tools and Methods Supporting Process Verification

One of the benefits of implementing BI technology (Fig. 5.7) is the possibility of not only verifying processes in the context of achieved performance (batch processing in historical mode).

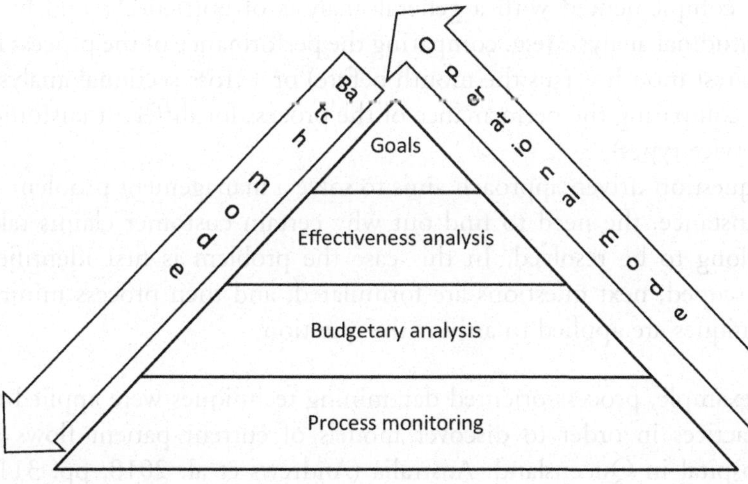

Fig. 5.7 The functional architecture of Business Intelligence Systems (*Source* Rostek 2010, p. 167)

Operating BI in real-time mode (ex ante) provides the ability to control process parameters and neutralize deviations before they significantly impact the final result. Such activities may prove useful in public administration, particularly in terms of monitoring the performance of public services (primary processes) but also with regard to supporting processes (logistics, supplies, procurement). In turn, once process performance is completed, it can be thoroughly analyzed to provide insights concerning its (Fig. 5.7): financial and quality efficiency, goal achievement, and process correctness. This very valuable information can later be used for the purposes of local optimization or global process improvement.

In-depth process analyses, facilitated by data mining tools that are increasingly included in BI system packages, are usually conducted in two distinct contexts (Dumas et al. 2013, p. 461):

- the exploratory approach - is driven by the curiosity to understand how a process is executed without any specific question at hand. In this approach, it typically starts by discovering a process model or by checking conformance against an existing model. This analysis is typically complemented with a general analysis of bottlenecks and by a longitudinal analysis (e.g. comparing the performance of the process in the latest month versus the month before) or a cross-sectional analysis (e.g. comparing the performance of the process for different customer or service types),
- the question-driven approach aims to solve a management problem—for instance, the need to find out why certain customer claims take too long to be resolved. In this case the problem is first identified and scoped, next questions are formulated, and then process mining techniques are applied to answer the question.

As an example, process-oriented datamining techniques were applied to real practices in order to discover models of current patient flows at the Hospital in Queensland, Australia (Andrews et al. 2018, pp. 311–333). This project demonstrated that process mining is applicable to complex, semi-structured processes like those encountered in the healthcare domain. A comparative process performance analysis yielded some insights into patient flows, including recognition of recurring data-quality issues in datasets extracted from hospital information systems. The templated recognition and resolution of such issues offers a research opportunity to develop a (semi-)automated data-cleaning approach that would alleviate the tedious manual effort required to produce high-quality logs. The project highlighted the importance of hospital information systems collecting both start and end times of activities for proper performance analysis (duration, wait time, bottlenecks). Additions to the process-mining toolset include novel comparative process-performance visualization techniques that highlight the similarities and differences among processes and their further iterations.

It is noteworthy that the presented example clearly shows how this type of in-depth analysis could not only answer the posed analytical question, but also provide a lot of valuable information about the state of the process, effectiveness of its execution, identifiable errors and executive bottlenecks. All in all, it is a very valuable and significant contribution to process verification as an element of the overall organization's management model.

One also should not forget the possibility of presenting analytical results in the form of interactive reports known as performance dashboards, which allows mining the analyses—from the most general level (the most aggregated form of the result) and all the way down to the highest detail (input data on which the result is based). Hence, the decisionmaker receives general, aggregated information indicating, for instance, the current condition of the task or project's budget (correct, borderline, exceeded), but on demand and without the need to request additional analyses or elaborative reports, all elements contributing to the result (intermediate results) as well as the input data are also immediately available. On the one hand, this prevents unnecessary information overload, while on the other, it saves time when additional supporting information is required.

In the case of public administration, the control and verification of process performance (especially in terms of primary processes aimed at the provision of public services) can successfully rely on workflow systems. Workflow is a concept described as process automation, wherein documents, information, or tasks are passed on from one operator to another in accordance with the adopted procedures regarding implementation and acceptance of subsequent elements of the flow. It is therefore a method of defining and controlling policies and channels of information exchange between respective process operators. Typical features of workflow systems that increase their viability in terms of process monitoring and verification include tools supporting:

- process definition – graphic or textual tools facilitating process definition, with the possibility of assigning personnel (operators), establishing rules to determine control thresholds and task performance progress, as well as dynamically amending processes and tasks assigned to respective operators,
- decision-making – rules are established for every stage of the process and its flows, which facilitates correct correlation of workflows and resources, control of the tasks' time-consumption and productivity, as well as easier scheduling and adequate process provisioning,
- task automation – workflow systems have the capacity for automatic initiation of subsequent tasks once the defined triggering conditions are met (e.g. upon acceptance of the previous task's performance or emergence of a specific deadline). This organizes the process flow while at the same time reducing the workload of decisionmakers by taking over the direct supervision of the same,
- executive worklists – allowing the process operator to quickly identify assigned tasks, with a particular focus on deadlines, priorities, as well as legal and quality requirements. Some systems also support workload and worktime forecasting. In this context, employees and decisionmakers are provided with a tool facilitating task performance monitoring, labor intensity, and productivity for each respective employee,
- process monitoring – workflow systems can monitor predefined key indicators for elements of process supervision and control, which allows them to fully and directly perform the function of process performance verification,
- simulation and prototyping – tools that can simulate workflows or create prototypes and deploy pilot versions of a given workflow. This offers a way of verifying prototypes in test environments before full

deployment, which can reduce costs and verify the quality of a newly implemented element,

- data management – workflow systems commonly feature a variety of tools facilitating data processing, including databases, analytical applications and reporting applications, spreadsheets, text editors, etc. Workflow systems can also be integrated with other systems operating in the organization, e.g. analytical and reporting applications, tools used for visualization and distribution of information, or corporate portals,
- dissemination, transfer, and distribution of documents – each person involved in the preparation, verification, or approval of a document is granted automatic access, which can be limited in terms of duration, scope (documents divided by competence levels), or access authorization and responsibility (read-only, editing, review, deletion).
- event notification – operators and decisionmakers are notified when a task consistent with the planned stages is initiated or when a task list performance is backlogged. This reduces response to planned events but also in terms of diagnostics and handling emerging threats,
- multichannel distribution and dissemination of information – the capabilities of workflow systems in this respect continue to be improved. Currently, most already support not only user applications but also corporate portals, internal networks, as well as mobile applications, which allows employees to access necessary information at anytime and anywhere,
- administration and security – with such a wide scope of applications, it is naturally paramount for such systems to be adequately protected against unauthorized access. Consequently, workflow systems feature security measures such as user authentication, authorization adjustable depending on the employee's needs and competence levels, as well as process continuity safeguards in the event of a temporary malfunction or system failure.

Process verification through ex post analysis is used with a view to process improvement in subsequent iterations. Nonetheless, one can expect to be able to verify the efficiency and effectiveness of the process even before its initial deployment. This is facilitated by simulation tools used when designing an operational model of a system and conducting experiments with this model for the purpose either of understanding the behavior of the system or of evaluating alternative strategies for the development or operation of the system (Thome 1993). They have to be able to reproduce selected aspects of system behavior modelled to an accepted degree of accuracy. The observed widespread incorporation of simulation methods in process management systems is understandable given their functionalities (Devjak et al. 2004, p. 2):

- recognition of conditions of the process execution,
- experimentation on the model before introducing the modifications into reality,
- forecasts of process development and changes of conditions during its execution,
- analyses of process realizations, the differences in comparison to the plan and the factors causing process modifications,
- execution of other management and controlling functions in the area of management of the process and its development.

Evaluation of public service provision in the context of the associated process covers three areas of assessment (Devjak et al. 2004, p. 3): process goal, process performance, and process efficiency. Each of these scopes can be modeled and evaluated with the use of simulation tools. A simulation model is a method that allows the prediction of process goals along with projections of the process's end results and parameters. According to this concept, the simulation method draws upon previous process iterations, which increases the applicability of the method in terms of exactness of process efficiency evaluation. In particular, a simulation model can provide (Devjak et al. 2004, p. 5):

- information on the duration of procedures and possible performance problems,
- ascription of qualifications and competencies needed to conduct the procedures, and prediction of procedural schedules,
- assessment of the suitability of equipment and manning,

- information on the persons responsible for irregularities identified in the procedures,
- economic evaluation of the procedures and development of a system of indicators and measures.

Said information and activities can fully facilitate not only ongoing verification of process performance but also long-term improvement of the same, while simultaneously requiring minimum investment of assets and resources. In this context, a simulation should be perceived as a method integrating the processes or research, development, design, and operation (Fig. 5.8).

Thus perceived, simulation methods become a significant tool usable over the entire process lifetime—from its inception and design, through implementation and verification, all the way to eventual improvement. Hence, one is recommended to use supporting tools that facilitate not only process modeling but also their simulation and optimization. WebSphere Advanced Business Modeler is one of such solutions as it combines functionalities enabling one to model (Pihi et al. 2010, p. 649): process structure and flow, organizational architecture and structure, data

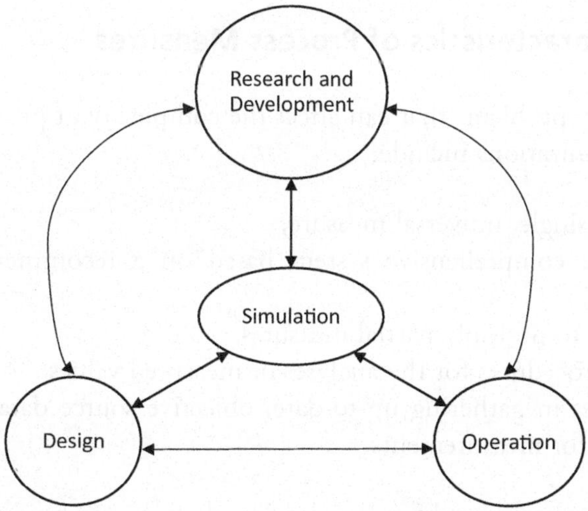

Fig. 5.8 *The new paradigm of Process Engineering* (*Source* Dimian 2014, p. 37)

structure, resource structure, time and location-wise ascription of process iterations, with the ability to generate process simulations and analyze their results. This integration of a wide range of functionalities provides the user with a comprehensive tool supporting the design, verification, and improvement of organizational processes. Notably, all potential improvements can first be tested in a virtual environment without the need for excessive financial and organizational outlays.

5.3 Comparison and Recommendations—Anna Kosieradzka and Katarzyna Rostek

In the case of production companies and public administration entities alike, process verification ought to duly account for applicable goal and management-oriented measures. A model to which organizations can relate when analyzing such determinants is provided by the KPI system whose characteristic feature is its uniqueness and customizability to the specificity of a given organization and its operations.

5.3.1 Characteristics of Process Measures

The primary problems that can affect the complexity of process verification in organizations include:

- lack of a single, universal measure,
- lack of a comprehensive system based on a recommended set of measures,
- tendency to multiply partial measures,
- lack of procedures for the analyses of measured values,
- difficulties in gathering up-to-date, objective source data to serve as the basis for measurements.

Measuring systems ought to be developed on a case-by-case basis and adapted to the specificity of a given organization and its operating objectives. Excessive expansion of system measures is ill-advised, particularly in terms of partial measures, as oftentimes efforts aimed at improving local performance do not guarantee a desirable global result. Moreover, the significance of measures is conditioned by their practical applicability, they should always be targeted, dedicated, and deliberately chosen. The more measures are introduced, the harder it becomes to meet those requirements.

The adopted system of measurement has a particularly strong impact on employees and their behavior, as observed by Goldratt: *Tell me, how you will measure me, and I will tell you how I will behave. If you measure me in an illogical way … do not complain about illogical behavior. (Goldratt 1990, p. 26).* The logic of this statement dictates that a measure should not only have a specific purpose and scope of applicability, but also be compatible with the organizational culture, so that it will motivate employees rather than incite pathological behavior. It seems that the latter aspect plays a particularly significant role in the context of public administration institutions where measures often relate directly to the method and quality of task performance, which means that their potential for positive influence significantly impacts the functional capacity of the entire organization.

5.3.2 Methods of Defining Process Measures

In the case of production companies, measures are provided by well-established theoretical and practical indicators pertinent to operations that control and monitor the key elements of production, auxiliary, and management processes. Therefore, the difficulty lies in selecting the optimum set of measures to best reflect the organization's objectives, rather than deciding on the procedure for their identification and definition.

In the case of public administration, one would be hard pressed to single out a standard set of measures that could be applied a priori in terms of selection and choice. The definition of measures is conducted by carefully analyzing each process and its individual elements relative to the goals, assumptions, requirements, and limitations at the level of an activity, process, or group of correlated processes. This entails progressing from specifics to generalizations, where at the particular level (a single activity, process, or procedure) specific characteristics are defined, which are then grouped, generalized, and selected at progressively higher levels of process management in such a way, so as to end up with a group of only a handful, up to a dozen or so, measures that are the best representation of the process at hand. Given the multifaceted nature of such an analysis, the activity is bound to be complex and require considerable accuracy, as well as experience and ability to perceive process elements in the context of the multidimensional, global system of the organization.

5.3.3 Tools and Methods Facilitating Process Verification

While identification and definition of new measures constitutes a creative activity that would be difficult to automate given the current state of knowledge, the verification of process performance results based on predefined measures as such is very much automatable. This can be accomplished with the use of analytical, simulation, and control tools capable of processing incoming results, also in real-time modes. Their application provides a wide spectrum of opportunities in terms of optimization, perfection, and quality improvement of processes, in some cases by way of fully automated measures performed in real time (e.g. process correction based on results obtained from the analysis of measures assigned by the system in real time, assuming full automation of the process activities and full integration of production and analytic systems). In the case of public administration institutions, the deployment of such solutions can enhance analyses by incorporating elements such as e.g. the extent of influence on the customers, consequences for the public image, or forecasts of non-typical and rare events. In this context, the tools and methods utilized prove particularly valuable in the areas of strategic decision making, which ultimately also has a bearing on operational optimization.

Apart from new technologies, there are also other comparative methods and techniques that can prove useful, for instance benchmarking which, particularly in the context of public administration, can help to verify activities against known, proven and practically verified models. Exchange of experiences with the simultaneous adaptation of verified solutions to new circumstances will, on the one hand, help to strengthen cross-institutional bonds and cooperation (which constitutes added value in itself), and on the other, facilitate further improvement of established models and solutions (which in turn contributes to the development of knowledge and science in the area of management methods, models, and tools). Therefore, in this age of automation, digitalization, and technological advances, one should not forget about the benefits of simple communication, exchange of information and knowledge, and various other forms of collaboration.

References

Andrews, R., Suriadi, S., Wynn, M., ter Hofstede, A. H., & Rothwell, S. (2018). Improving Patient Flows at St. Andrew's War Memorial Hospital's Emergency Department Through Process Mining. In *Business Process Management Cases* (pp. 311–333). Cham: Springer.

Devjak, S., & Peček, B. (2004). Simulation Techniques in Public Administration Procedures. *Economia Aziendale Online, 1*(2004), 1–17.

Dietrich, E., & Schulze, A. (1999). *Statistical Procedures for Machine and Process Qualification*. Milwaukee: ASQ Quality Press.

Dimian, A. C., Bildea, C. S., & Kiss, A. A. (2014). Introduction in Process Simulation. In *Computer Aided Chemical Engineering*, Vol. 35, pp. 35–71. Elsevier.

Dumas, M., La Rosa, M., Mendling, J., & Reijers, H. A. (2013). *Business Process Management*. Berlin: Springer-Verlag.

Goldratt, E. M. (1990). *The Haystack Syndrome: Sifting Information Out of the Data Ocean*. Croton-on-Hudson, NY: North River Press.

Harrington, H. J. (1991). *Business Process Improvement*. Pproductivity and Competetiveness, McGraw-Hill, New York: The Breakthrough Strategy for Total Quality.

ISO 9000:2015. Quality management systems—Fundamentals and vocabulary.

Kaplan, R. S., & Cooper, R. (1998). *Cost & Effect: Using Integrated Cost Systems to Drive Profitability and Performance*. Boston: Harvard Business School Press.

Kaydos, W. (1999). *Operational Performance Management*. Boca Raton: CRC Press LLC.

Kosieradzka, A. (2012). *Zarządzanie Produktywnością w Przedsiębiorstwie (Productivity Management in the Entrprise).* Warszawa: C. H Beck.

Molenda, M. (2016). Zasady Ustalania Mierników Procesów w Systemie Zarządzania Jakością. *Systemy Wspomagania W Inżynierii Produkcji, Nr, 2*(14), 249–257.

Nakajima, S. (1988). *Introduction to TPM*. Portland: Productivity Press.

Oakland, J. S., & Oakland, R. J. (2019). *Statistical Process Control*. London and New York: Routledge.

Pande, P. S., Neuman, R. P., & Cavanagh, R. R. (2000). *The Six Sigma Way: How GE, Motorola, and Other Top Companies are Honing Their Performance*. New York: McGraw-Hill Professional.

Pihir, I., Zajdela, N., Dušak, V. (2010). Survey of Simulation Capabilities of the IBM WebSphere Business Modeler Business Process Modeling Tool on the Example of Processing a Loan Application. *Proceedings of the International Conference on Information Technology Interfaces*, ITI. pp. 645–650.

Rostek, K. (2010). Business Intelligence for SME. In E. Lechman (Ed.), *SMEs and Enterpreneurship* vol. II, Gdańsk University of Technology Publishing House, Gdańsk, pp. 164–190.

Rummler, G. A., & Brache, A. P. (1995). *Improving Performance*. San Francisco: Jossey-Bass Publishers.

Thome, B. (Ed.). (1993). *Principles and Practice of Computer-based Systems Engineering*. Chichester: Wiley Series in Software Based Systems. John Wiley & Sons.

6

Process Perfection

Dr. Anna Kosieradzka and Katarzyna Rostek

6.1 Process Perfection in Production Companies—Anna Kosieradzka

The following section will present the theoretical and practical aspects of process development and improvement in organizations, on selected examples of production companies. The presented content is based on the results of our own and secondary studies.

6.1.1 Classical Approach to Production Processes Improvement

Improvement of production processes has been treated as an inherent part of the same for over a century. The trend towards perfecting production processes was initiated in the USA as part of the so-called classical school of organization and management aiming to develop methods and techniques that would facilitate higher work performance. Some of the best-known representatives of the classical management perspective

A. Kosieradzka and K. Rostek, *Process Management and Organizational Process Maturity*,
https://doi.org/10.1007/978-3-030-66800-6_6

include (Griffin 2017, p. 41): Frederic Taylor, Lillian I Frank Gilbreth, Henry Gantt, Harrington Emerson, and Henry Ford. A French engineer and chemist, Henry Le Chatelier, developed a universal approach to gradual improvement of work processes, nowadays known as the organizational cycle, which was composed of five distinct stages:

- choice of the goal to be accomplished,
- analysis of resources and conditions necessary for the goal to be achieved,
- gathering of the necessary resources,
- execution of the adopted plan,
- control of the expenditures.

Based on the works by Taylor and Le Chatelier, a diagnostic approach to improvement was developed, as presented in Fig. 6.1.

In the early twentieth century, many new methods were developed with a view to improving processes, particularly repeatable ones, e.g. (Kanawaty 1992):

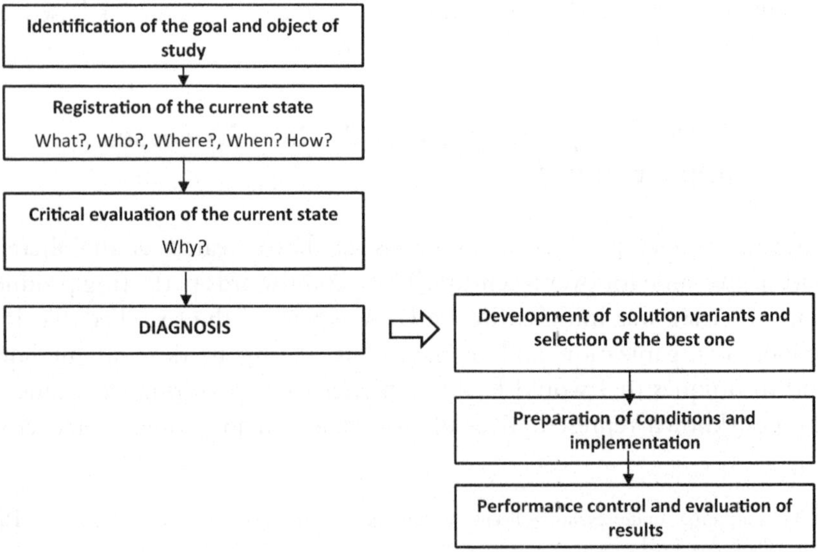

Fig. 6.1 Diagnostic approach to improvement (*Source* Own elaboration)

- method study (movement of workers, micromotion study, two-handed process chart),
- work measurement (time study, work sampling),
- movement of workers (string diagram, worker-type flow process chart),
- movement of material, string diagram, flow process chart).

Some of the methods continue to be used today, e.g. string diagrams also known as spaghetti diagrams, work sampling, and process flow charts identifying waste and nonvalue-adding activities. New life was breathed into those methods by Japanese researchers who included them among the problem-solving tools utilized under Kaizen and Lean Management concepts.

6.1.2 Contemporary Approach to Production Process Improvement

The foundations for contemporary production process perfection were derived from a paper by Nadler (1967), who posited that rather than seek improvement variants for the respective elements of a process, one should aim to develop a comprehensively new concept of the same. This is a so-called "calibration" approach which entails starting from the ideal system and gradually moving towards a solution feasible under the given limitations. The ideal system is a non-existent, theoretical formulation (not unlike infinity in mathematics) that can serve as a specific system of reference facilitating uninhibited innovativeness in the development of new solutions. These exact notions can be traced in the concept of Reengineering defined as *fundamental rethinking and radical redesign of business processes to achieve dramatic improvements in critical contemporary modern measures of performance, such as cost, quality, service, and speed* (Hammer and Champy 1993, pp. 90–91).

Reengineering is a concept best reflecting the trend towards process improvement in all types of organizations, but as far as production companies go, they continue to rely primarily on the Kaizen concept—inherently related to the Toyota Production System and constituting the basis for Lean Management. In the foreword to the 2007 Polish edition of his book "*Kaizen. The Key to Japan's Competitive Success*", Maasaaki Imai wrote: "*While western managers prefer continuous changes based on technologies and significant investments, their Japanese counterparts opt to improve processes in small steps, based on common sense and employee engagement. The western approach is nowadays referred to as innovativeness. In order to survive in today's competitive environment, managers must urgently adopt the principles of both innovativeness and Kaizen.*"

It is hard not to notice that Kaizen is in fact a contemporary version of the diagnostic approach, while what Masaaki Imai refers to as innovativeness is rather close to the core idea of reengineering. His suggestion that both of the same ought to be adopted in a company, as only by synergizing innovativeness with Kaizen can one create conditions for lasting success, simultaneously explains the reasons of so many unsuccessful implementations that were overly radical and neglected to consider the human factor. To emphasize the differences between implementations of radical process reengineering and continuous "small steps" process improvement, the characteristic features of the concepts are presented in Table 6.1.

For well over two decades, a skillful combination of Kaizen and innovations has been the method of choice for companies worldwide aiming to perfect production processes. It remains to be seen for how long this trend can continue and how the Fourth Industrial Revolution that is now unfolding before our eyes will ultimately impact the balance between innovative technological breakthroughs and team-based day-to-day problem solving.

Klaus Schwab (2016, p. 2) wrote: "*While the profound uncertainty surrounding the development and adoption of emerging technologies means that we do not yet know how the transformations driven by this industrial revolution wild unfold, their complexity and interconnectedness imply that all stakeholders of global society[...] have a responsibility to work together to better understand the emerging trends.*" Technological progress has always had the effect of destroying some workplaces, sometimes replacing them with others, related to doing something new, somewhere else (Schwab 2016, p. 36).

Table 6.1 Kaizen and Innovations—a comparison

No.	Feature	Kaizen	Innovations
1	**Result**	Long-term, permanent, gentle	Short-term, radical
2	**Pace**	Small steps	Large leaps
3	**Timeframe**	Continuous and expanding	Project with a defined start and finish
4	**Change**	Gradual and continuous	Sudden and transitory
5	**Participation**	Everyone	Project team—several handpicked "masters"
6	**Approach**	Collective, group effort, systemic approach	Strong individualism, individual ideas and efforts
7	**Conduct**	Maintaining and improving on the status quo	Demolition of the current status quo and its in-depth reengineering
8	**Idea**	Conventional knowledge, traditional technologies and accepted code of conduct	Technological breakthrough, new inventions, new theories
9	**Practical requirements**	Moderate investment but large effort required to solidify	Large investment but less effort required
10	**Orientation**	People	Technology
11	**Evaluation criteria**	Local measures focusing on partial processes	Global measures (Quality, Cost, Delivery)

Source Own elaboration based on Imai (1986, p. 24)

Various categories of work, particularly ones that entail mechanical repeatability, already have been or soon will be automated, hence the focus of process improvement will naturally shift from performance to planning. The same also pertains to non-production processes—in services or administration.

Apart from Kaizen and reengineering, another contemporary concept commonly applied to process improvement is known as Six Sigma. The three key methodologies will be discussed in greater detail below.

6.1.3 Kaizen

As already mentioned, the core of the Kaizen approach lies in continuous process perfection through small improvements introduced by all members of the workforce. Kaizen changes are introduced in three areas differing in terms of their level, scope, and complexity:

- Management oriented Kaizen,
- Team oriented Kaizen, and
- Individually oriented Kaizen.

Management oriented Kaizen focuses on undertakings of strategic significance. Its aim is to restructure processes, systems, and procedures at the level of the entire organization. Problems of this type are delegated to dedicated project teams and task groups composed of staff members (mainly managers and specialists) from various parts of the organization, as well as external consultants. The object of such efforts is to develop new systems, technical improvements, and innovative solutions, which often tends to require considerable financial investments. Management oriented Kaizen entails creation of task groups to conduct the planned activities, which should be managed with the use of rules, methods, and tools applicable to project management. Undertakings falling within this category are rather indistinguishable from reengineering of primary, supporting, and management processes.

Team oriented Kaizen focuses on repeatable small group activities conducted by teams with a variety of designations (e.g. quality circles, productivity improvement circles, fault-free operation circles, work safety groups, etc.). Such groups should emerge autonomously, i.e. on the initiative of employees themselves. Members of small groups are typically rank-and-file workers from a single organizational unit, with group leadership falling to one of the members, not necessarily the formal manager.

The autonomous character of small groups is also expressed by the fact that the members determine their own areas of activity and improvement objectives. This type of Kaizen—team-based problem solving—is the most important element of the Kaizen program in production companies.

Individually oriented Kaizen takes the form of requests submitted individually by employees based on the assumption that the potential work improvement at the levels of a single person is virtually unlimited. Individual improvement ideas (so-called suggestions) should emerge almost immediately after the initiation of the Kaizen program as they tend to be primarily organizational and require no financial outlays, mostly entailing change of a given employee's attitude to their work.

Attempts to use Kaizen concepts in everyday activities of organizations have been made in numerous countries, however, Toyota factories in Japan remain the unsurpassable model for the practical utilization of employees' knowledge, experience, and ingenuity. As follows from our research, problems with Kaizen implementations stem primarily from the fact that many managers fail to take rank-and-file employees' suggestions seriously, and that the philosophy of continuous improvement remains insufficiently "furnished" with actual organizational solutions. Small groups do not receive the necessary support within corporate organizational structures, there are no systems to facilitate submission, evaluation, reward, or adequate support for implementation processes. There are also insufficient systems dedicated to incentivizing employee suggestions and communicating their accomplishments (Fig. 6.2).

In order to improve a process, one needs to identify **(I)** existing problems and design solutions to facilitate their elimination. Team-based problem solving in small groups (team-oriented Kaizen) usually follows the pattern described in the so called Deming's PDCA cycle (*Plan, Do, Check, Act*) considered to be one of the key approaches to problem solving in general (Fig. 6.3).

Before any improvement is implemented, its goal must be specified along with a realistic plan for its introduction complete with a specification of tools and methods involved. This is the planning (P) stage which entails:

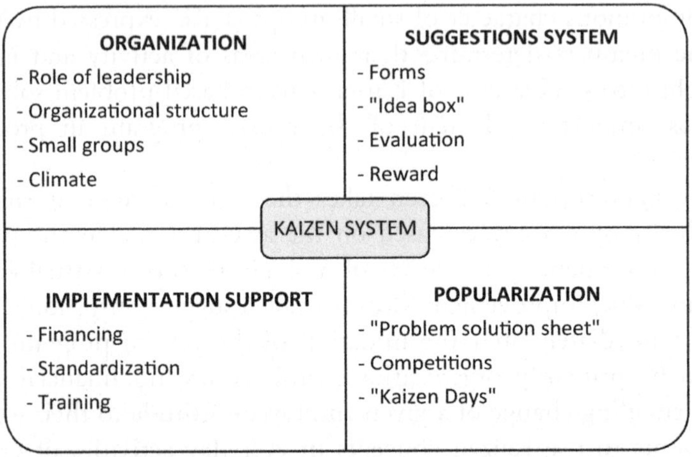

Fig. 6.2 Elements of the Kaizen system in an enterprise (*Source* Own elaboration)

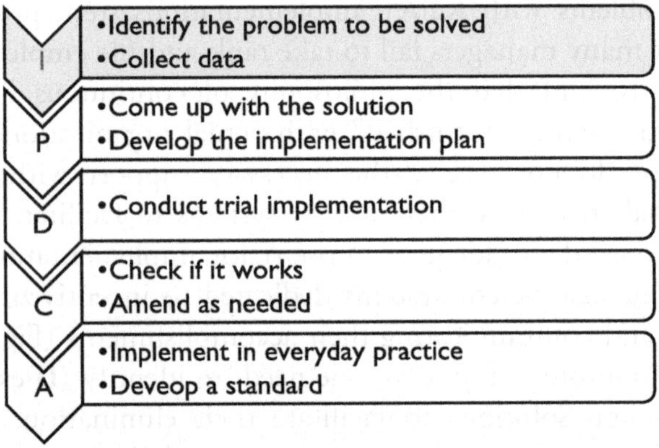

Fig. 6.3 Problem solving scheme (*Source* Own elaboration)

- formulating the problem that needs to be solved,
- agreeing on the action plan with the group members,
- solving the problem:
  - gathering data,
  - analyzing data,

- • collecting improvement ideas,
- • choosing the best idea,

– assigning tasks to respective group members.

The next step is implementation of the solution (D), i.e. putting the developed plan in action to verify the correct choice of tools and methods. Once the solution has been implemented, its effectiveness is observed in everyday practice (gathering and evaluating results) to verify (C) whether any problems or difficulties can be identified and consequently require corrective action. If any inconsistencies are observed between the results and goals, the so-called perfection in day-to-day operation (A) takes place.

Stabilization of the solution requires standardization. The significance of the same as one of the key factors conditioning successful implementation of processes was already emphasized in chapter 4.1.4. After every amendment to a process, it is important to duly update the affected work standard, as such changes run the risk of increasing process variability which can be counteracted by developing detailed standards for any new activities involved. Further improvement will only be possible once the process has been adequately stabilized.

Process reengineering

As defined by M. Hammer and J. Champy, reengineering of production processes involves in-depth reimagining and redesign of the same with a view to inducing a "dramatic" performance improvement in terms of key determinants of competitiveness, i.e. quality (Q), cost (C), delivery terms (D), and speed of response to customer needs (S).

Initially, BPR was driven by Information technology (IT). Explosive advances in IT enabled the dissemination, analysis and use of information from and to customers and suppliers, as well as within enterprises, in new ways and in time frames that impact process, organization, design and strategic competences. Computer networks, open systems, client-server architecture, groupwork and electronic data interchange have opened up the possibilities for integrated automation of business processes (Oakland 2014, p. 229).

Various authors have suggested different definitions of steps involved in BPR. Some, including a number of consulting companies, have developed their own methodologies that emphasized the importance of specific measures. Nonetheless, many common elements are included in all of them:

- selection of the process(es) to be reengineered,
- establishment of a team responsible for implementing the changes,
- analysis and understanding of the current process,
- reengineering the process,
- implementation of the improved process,
- measurement of results (KPIs—Key Performance Indicators).

Some authors drew attention to such elements as:

- the need to verify the adopted reengineering goals to ensure that the restructured processes do not negatively affect other associated processes,
- the need to rain employees and prepare them for the adoption of the solution,
- the need to make efforts aimed at implementing the developed solution on a larger scale.

In production companies, radical process reengineering most often occurs in the following areas:

- new product design,
- implementation of new production technologies,
- automation and robotization of production processes, including transport and warehousing activities, material handling, machine retooling, etc.,
- introduction of quality control in production processes through the implementation of automated measuring systems combined with statistical process control (SPC),

- reorganization of processes by eliminating organizational barriers and integrating subprocesses previously performed by separate units (Sujova et al. 2019),
- computerization of production planning and control, with the reservation that simply introducing computerized systems in production control allows automatization of only the existing processes, it should therefore be preceded by cross-functional process integration.

Verification of the developed solutions often involves the use of simulation models facilitated by specialized software (e.g. Tecnomatix Plant Simulation by Siemens Industry Software or FlexSim by InterMarium).

New challenges for process reengineering are posed by technologies developed as part of the so-called Fourth Industrial Revolution, particularly IoT—Internet of Things, which aims to bridge the divide between physical and digital systems, as well as robotics, autonomous vehicles, drones, artificial intelligence, energy storage, new materials, or 3D print—along with the entirely new business models they bring about. This, however, is the probable song of the future. What then are the challenges right now?

The greatest of the same invariably remains ensuring that the implementation of a radical change in production processes is met with acceptance by the relevant stakeholders. The success of a BPR initiative in any organization, not only a production company, depends primarily on people employed at all levels of the hierarchy, as their resistance may lead to failure of the endeavor. It falls to the managers to create an organizational structure ready to welcome changes, even radical ones. While an atmosphere of competitiveness ("survival of the fittest") does not bode well for the success of a radical reengineering initiative, the experience of Japanese companies as well as their imitators worldwide points to the viability of the Kaizen concept which alternates between innovation and engagement of the entire workforce in the process of continuously perfecting the implemented solutions. Oakland emphasizes the need to improve on a BPR solution, particularly in the initial period of its functioning when errors and defects are most likely to emerge (2014, p. 136). He also notes that "*TQM ideals and Concepts provide a perfect platform for*

BPR projects and the creation of a learning organization capable of adopting to a radical changing environment" (Ibid., p. 242).

Six Sigma

Six Sigma originated from the TQM concept. Nowadays, it is often regarded as an off-road management concept and defined as a comprehensive and flexible program designed to reduce defects, lower costs, save time, and improve customer satisfaction. Six Sigma is a comprehensive system—a strategy, a discipline, and a set of tools—for achiving and susutaining business success. It is a strategy because it focuses on total customer satisfaction. It is a discipline because it follows a formal improvement model called DMAIC, and it is a toolset because it provides a formal set of statistical tools (Heizer and Render 2011, p. 227).

Six Sigma is described as best-practice approach to optimizing processes through actions aimed at meeting customer requirements and reducing special and common causes of process variability—Fig. 6.4.

Six Sigma assumes a continuous improvement of quality, which translates to the organization's profit by increasing customer satisfaction. The measure of the process quality level is the sigma value callculated as the number of defects per million opportunities—DPMO. Respective sigma quality levels are defined depending on the DPMO (Table 6.2).

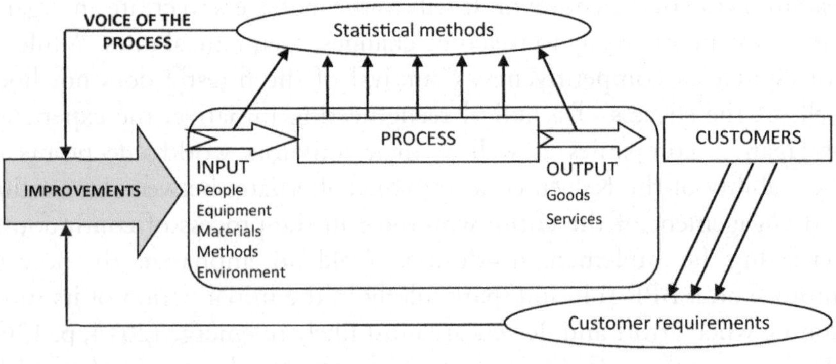

Fig. 6.4 Six Sigma Idea (*Source* Own elaboration)

Table 6.2 Sigma quality levels depending on DPMO

Indicator	Sigma level					
	1	2	3	4	5	6
Acceptable error number [DPMO]	697,700	308,537	66,807	6210	233	3.4
Share of products that meet the requirements [%]	30.9	69.2	93.3	99.4	99.98	99.9997

Source Gordon (2010, p. 25)

"Sigma level" refers to the standard deviation in the population. If the distribution of a given characteristic is described by normal distribution, then the sigma level determines the number of times the standard deviation of a given feature falls at the middle of the tolerance field (Gordon 2010, p. 26). The level "6 sigma" means that in the context of a given feature, only 3.4 errors will occur per million events—for instance a production process where only 3.4 deficiencies are identified per one million parts produced.

There are five fundamental phases or stages to applying the six-sigma approach with a view to improving process performance: Define, Measure, Analyze, Improve, and Control (DMAIC). DMAIC provides a breakthrough strategy and disciplined methods of using rigorous data gathering and statistically based analysis to identify sources of errors and ways of eliminating them (Oakland and Oakland 2019, p. 344). Activities in the relevant phases are shown in Table 6.3.

Presented below is a case study illustrating production process improvemet in accordance with the Six Sigma approach (Kosieradzka and Ciechańska 2018). The described example comes from an enterprise manufacturing industrial automation devices. The company operates globally offering products in four market segments: cooling, heating, drives, and power solutions. The presented case study pertains to the comoany's factory in Poland, supplying the refrigeration and cooling segment. The factory is divided into 5 production lines where c.a. 900

Table 6.3 Six Sigma phases—DMAIC methodology

Phase	Description	Steps	Tools	Mission
D	***Defining the problem and launching a project leading to a solution*** Developing a plan for improving the company's performance	Business analysis of the project (goals and critical points) Booking resources with the sponsor Setting up a team (roles and responsibilities) Work schedule Identification and description of processes Defining the customer's voice (requirements)	• Project Charter • SIPOC • CTQ Matrix • Stakeholder Analysis	The project is defined The current state and target state are depicted and the process to be improved is market off Customer and business requirements are clearly defined
M	***Measurement of process performance*** It is based on measuring the company's systems and processes—measuring the current state, setting goals	Process map (general and detailed) Identification and selection of variables Data collection plan Evaluation of the measuring system (measuring instruments, type of measurement, recording) Initial process capability (determination of non-compliance level, DPMO, sigma level)	• Measurement Matrix • Operational Definition • Measurement System Analysis • Sample Size and Strategy • Charts and Diagrams	The starting situation is captured Key figures and operational definitions are developed, the measurement system analysis is completed and the data collected

Phase	Description	Steps	Tools	Mission
A	***Data analysis and identification of improvement opportunities*** Processing the data collected from the previous phase and comparing them to certain patterns, objectives or established initial process abilities	Analysis of correlations between variables Verification of hypotheses Elaboration on a list of root causes of the problem—identification of sources of variation	• Cause & Effect Diagram • FMEA • Process Analysis • Value Sream Map • Hypotesis Tests • Regression • DOE	The causes for the problem are identified All possible causes are collected and summarized inti the decisive key figures through process and data analysis
I	***Process improvement*** Introducing a change that will improve the current quality status—it can be the improvement of the process or the business management system	Generation of solutions Evaluation and choice of solution Risk assessment Action plan Implementation of the solution	• Brainstorming • "Must" Criteria • Effort-Benefit Matrix • Criteria-based Selection • Piloting • Roll-out Planning	The solution is implemented Possible solutions are generated on the basis of core causes, systematically selected, and prepared for implementation

(continued)

Table 6.3 (continued)

Phase	Description	Steps	Tools	Mission
C	***Control*** Monitoring the key process features in the enterprise to provide the intended capabilities, specific values of key variables and standardization of best practices	Evaluation of the results obtained Setting a new standard Creation of a control plan Assessment of business results Closing the project	• Documentation • Procedural Instructions • Control and Run Charts • Reaction Plan and Process • Management Diagrams	The sustainability of the results is secured The implemented solutions are documented and will be monitored using key figures A reaction plan secures prompt intervention

Source Own elaboration based on Shankar (2009, pp. 16–17) and Lunau (2008, p. 12) operators are employed. The data used in the subsequent parts of this chapter were gathered for the process of manufacturing water regulating valves.

The improvement efforts were motivated by production process unstability manifested by the lack of repeatability in obtaining key characteristics during the production of element A, to be included in product X—a kind of a switch. This led to the loss of functionality by product X—it did not operate as required by the customer. This signalled th eneed for a management review of the specifications and requirements pertinent to the production of element A. In order to solve the problem using the DMAIC methodology, an interdisciplinary team was established with the participation of R&D engineers. The team took action as described below.

Define. The project began with defining and verifying requirements for element A and product X. Based on the the detailed SIPOCR process map (Supplier, Inputs, Process, Output, Customer, Requirements), input and output variables of the process were defined and their mutual influence was determined. The key parameter of the process was the weight of element A which was referred to the requirements from the product specification (defined weight tolerances for element A). In the existing process, upon initiation of each element A production cycle, the first three pieces from the production batch were controlled. If the first three pieces of the order met the given tolerances, only one in 20 pieces would be controlled afterwrds. However, upon closer examination it was observed that the weight of many elements fell outside the tolerance range and was not repeatable. Moreover, when such faulty elements were used in product X, the latter completely lost its required functionality.

Measure. In the measurement phase, a plan for data collection and control of element A was developed. Based on the collected and recorded data, the process waste was verified (using the Pareto-Lorenzo chart) and referred to the quality standards for this area. A representative sample of 30 measurements was selected from register entries and analyzed using the Minitab statistical program.

Analyse. Based on the collected data, the process capability was analyzed. It was noticed that the weight of element A in the production process was not repeatable and the process was unstable (the values of C_p and P_p and C_{pk} and P_{pk} were not similar). A process can be said to be stable when the values of short-term (C_p and C_{pk}) and long-term indicators (P_p and P_{pk}) are similar. In addition, some values pertinent to the 30 samples were outside the tolerance limits. Given the above, steps were taken to determine whether the source of process variability resulted from special causes or whether it was due to common reasons. Based on existing process maps, the team created and analysed a C & E matrix (Cause and Effects Matrix) and determined the underlying source of the problem: 96.5% of elements were within the tolerance range but 3.5% fell outside the same in terms of weight.

A control card was created by adopting the tolerance limits as the upper and lower control limits. A clear trend of several consecutive events was clearly visible above and below the expected process values, which indicated the existence of a special reason for variability. The use of the existing PFMEA and a detailed map of the SIPOCR process helped to identify the root cause and relate it, with high probability, to the supplier of the raw material. The weight of component A was increased due to contamination resulting from the supplier's treatment method, which ultimately affected the operation of product X.

Improve. To confirm the hypothesis about the effect of impurities on the weight of element A, in the improvement phase, special samples (free of impurities) were prepared. The analysis of the data showed that the process complied with the requirements and the characteristics of element A were within the tolerance range after the elimination of the special reasons. In addition, it was observed that the functionality of product X was restored. A complaint procedure in accordance with the 8D process was initiated with the the supplier to avoid such problems in the future.

Control. The last step within the DMAIC cycle entailed assessment of the ability to meet the requirements on the 6σ scale. After eliminating the cause of variation, elment A reached a normal distribution and the DPMO dropped to 0. The level of short- and long-term capability was determined for element A ($C_p = 3.79$, $P_p = 3.32$). It was decided that C_p and $P_p > 3.0$ are satisfactory. Establishing a satisfactory level of process capability was only the starting point for further quality improvement. The original frequency of element weight monitoring turned out to be insufficient. Therefore, it was decided to introduce SPC cards based on upper and lower control limits (LCL = 19.6, UCL = 20.2, 6 sigma level (6 * standard deviation from the nominal value) and upper and lower specification limits (USL = 20.5, LSL = 19.3). The control plan for element A was updated by introducing weighting of every tenth piece in the process. In this way, in the event of a variation in the process, the operators observing the entries on the SPC card would be able to sqiftly inform persons responsible for the process. This improvement icreased the capacity to more effectively eliminate future sources of variability. Data collected with SPC cards can serve as the starting point for further improvements. It allows identification of special causes of variability and facilitates data collection. The results of the Six Sigma project are shown in Table 6.4.

Proper preparation of the company, both in the area of implementation of Six Sigma tools and in other process areas was the key to successfully completing the Six Sigma project in the discussed industrial automation enterprise.

6.2 Process Improvement in Public Administration—Katarzyna Rostek

As regards process improvement in public administration institutions, one can clearly observe prevalence of an adaptive approach that draws on solutions developed in the context of organizational, production, or project management. Models and good practices are also derived from the general administrative milieu which tends to vary in terms of process maturity as well as personnel experience and operating performance.

6.2.1 Public Administration Context

In much the same was that the concept of New Public Management serves as the premise for the implementation of process management, the concept of Lean Government (LG) constitutes the starting point for process improvement in public administration. Lean Government

Table 6.4 Case Study—before and after comparison for the Six Sigma process improvement project

Before implementation of SS tools	After implementation of SS tools
Central line: 19.9 g Tollerances: 19.4 – 20.4 g	Central line: 19.9 g Tollerances: LSL = 19.3 g USL = 20.5 g LCL = 19.6 g UCL = 20.2 g
Control chart 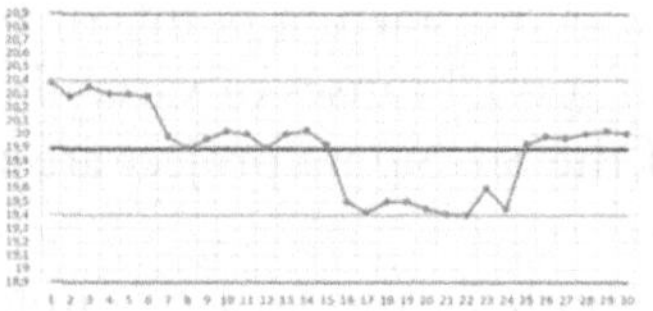	Control chart 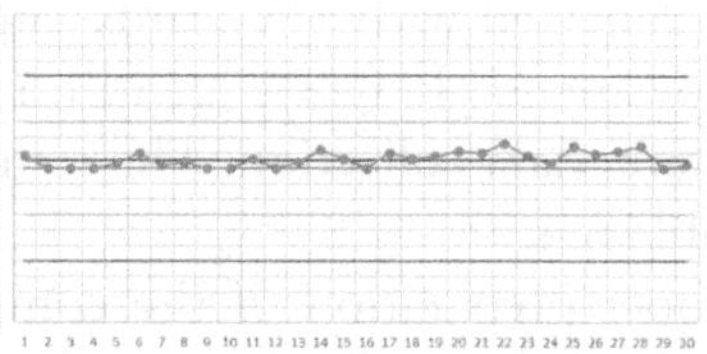
DPMO = 33333,00 Sigma level = 3	DPMO = 0 Sigma level = 6
μ = 19.89	μ = 19.96
Process capacity indicators Cp = 1.91, Pp = 0.53, Cpk = 1.88, Ppk = 0.52 Cp ≠ Pp , Cpk ≠ Ppk p < 0.005	Process capacity indicators Cp = 3.79, Pp = 3.32, Cpk = 3.34, Ppk = 2.29 Cp ≈ Pp , Cpk ≈ Ppk p = 0.028
Process capacity chart 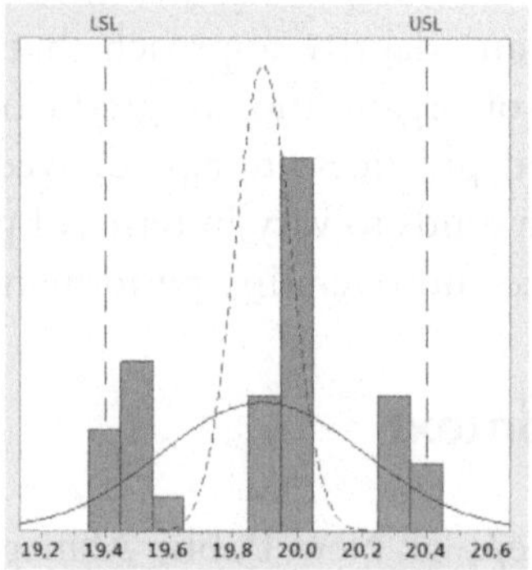	Process capacity chart 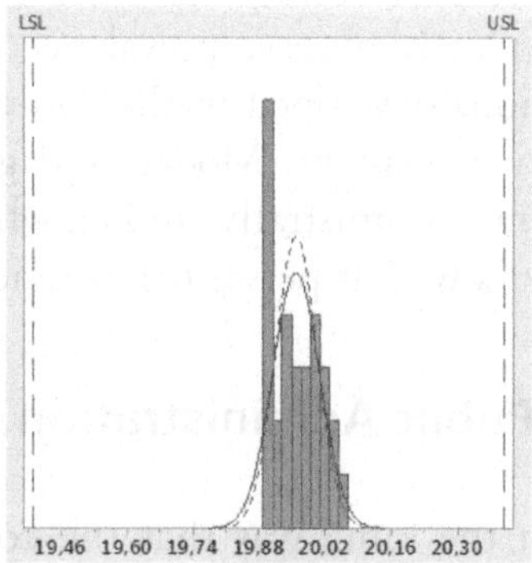

Source Kosieradzka and Ciechańska (2018)

focuses on governance and providing citizens (customers) with public services while continuously improving these functions by cutting out waste and inefficiencies in the processes. By identifying and eliminating unnecessary process steps and activities, it is easier to understand how the processes work, help make adjustments to optimize desired outcomes, and ensure that the processes functions smoothly and consistently. Lean government is a specific mindset and a discipline focused on increasing the capacity to do more and do better by:

- adopting a customer-service perspective that seeks to optimize value delivered to the public, the community, and other stakeholders,
- involving employees and external stakeholders in continual improvements and problem-solving activities,
- deploying a rapid continuous improvement framework that emphasizes implementation over prolonged planning,
- seeking to reduce the complexity of processes and variation in process outputs,
- using performance metrics and visual controls to provide rapid feedback to improve real-time decision-making and problem-solving,
- approaching improvement activities with a systems-focused mindset.

In order for the above to be feasible, a number of conditions must be met:

- be clear about purposes and bottom lines,
- know what customers want and what they value or need,
- build great (useful, reasonable) services,
- find a way to make the services better, faster and cheaper.

In turn, the above conditions cannot be satisfied without first thoroughly analyzing the existing state of affairs. The control and evaluation measures discussed in the previous chapter will certainly prove useful, if not downright necessary, in this context. However, such measures usually focus on the results of processes and the particular activities included therein, but rarely provide a broader perspective of the organization and its environment. It is therefore recommendable that the available self-assessment methods also be utilized to facilitate evaluation from the perspective of the entire organization and its milieu, naturally also with due consideration for the observed results of respective processes. One of the tools most commonly employed in public administration and used for the purposes of multidimensional evaluation of organizations is CAF—Common Assessment Framework. However, as its applications go beyond the thematic scope of this section, its detailed characteristics will be provided in the subsequent chapter of the monographic study.

6.2.2 Classic Approach to Process Perfection

The improvement actions should be implemented in accordance with the established method, which comprises the following phases (Aguirre et al. 2012, pp. 29–30):

- Phase 1. Project Understanding—the goal of this phase is to gain consensus over the problem to be solved, the scope of the project and the desired goals as stated in terms of the business process indicators,
- Phase 2. Process Analysis (as-is)—in this phase, the status quo is determined, key performance indicators are defined and analyzed, and areas for improvement are identified,
- Phase 3. Process Redesign (to-be)—this phase is intended to develop and simulate the corresponding process improvement alternatives,
- Phase 4. Change implementation—the goal of the implementation phase is to put the amendments in question in operation by changing procedures, job descriptions and work assignments.

The above procedure is to serve a single overriding purpose—to identify elements that require correction and determine the most effective way of implementing the necessary changes. The performance of the thus defined task is possible either by establishing a dedicated and experienced project team, or by conducting a benchmarking analysis. Benchmarking involves a comparison of the organization's processes, products, services, or practices with their counterparts observed in leading organizations on the market—sometimes referred to as benchmarking partners. The following stages of a benchmarking analysis can be identified (Jackiewicz and Jackiewicz 2017):

- identifying the area where a functional improvement is expected to occur,
- determining the characteristics that will be subject to analysis,
- identifying partners (benchmarking group),
- securing the benchmarking partners' consent for the collaboration,
- appointing people responsible for performing the study and processing its results,
- identifying and analyzing deviations from the best results in the group,
- planning and implementing changes aimed at minimizing the deviations and moving closer to the best results in the group,
- cycling back to the beginning of the process and identifying new area for continuous improvement in the organization.

Should we treat the benchmarking process as an analytical project, we could single out four distinct phases of the same (Table 6.5).

During the planning phase, the most important step entails identification of the subject matter which can be a particular process, part

Table 6.5 Benchmarking project phases

Project phases	Process stages
Planning	– identification of the subject matter of study – identification of comparable organizations – choice of data collection methods
Analysis	– conduct of the study – identification of efficiency deviations – determination of levels expected in the future
Integration	– dissemination of benchmarking results – identification of the functional goals of changes
Implementation	– preparation of the change implementation plan – initiation of implementation activities – implementation monitoring – identification of new benchmarks

Source Bogan and English (2006)

thereof, or specific activity. To facilitate comparison of results obtained in different organizations, it is also necessary to identify quantitative and qualitative process measures. These will be used to determine the extent of discrepancies between the respective benchmarking partners.

The analysis phase involves the performance of a comparative study and identification of qualitative and quantitative discrepancies between the results obtained for respective organizations. Information on the existing deviations will be used in the determination of value levels expected in the future. This stage can also generate proposals for temporary measures that will be easy to adopt, cheap to implement, and capable of yielding a considerable, if short-lasting, changes in respective areas. The purpose of the same is to temporarily improve the organization's functional capacity until strategic solutions are ready to be employed. Strategic measures are, by principle, long-term and targeted, but usually involve investment and deeper organizational changes. For this reason, their introduction tends to require wider timeframes and full project-wide implementation.

The purpose of the integration phase is to synthetize the results of the analysis phase and provide the same to decision makers who then determine the extent of necessary changes in the organization. The implementation phase puts the determination of the integration phase in action and proceeds to identify subsequent elements to undergo comparison and control.

When applied in accordance with the described approach, benchmarking becomes a method of continuous improvement and development of processes in an organization.

6.2.3 Agile Approach to Process Perfection

Similarly to the methods applicable in process implementation, one of the most critical obstacles and limitations faced in the context of process perfection relates to the time of identifying the problem, developing a solution, and ultimately implementing a change. Especially in large institutions, where any process change is likely to necessitate adequate organizational changes, the longer the time needed to redesign a process, the higher the cost and the more severe the consequences faced at the management level. Hence, again not unlike in the case of process implementation methods, also here agile solutions capable of reducing the timeframe and increasing the solution's flexibility are highly welcome. One example of such a method is the rapid process redesign methodology known as the NESTT, developed at the Queensland University of Technology and facilitating accelerated process improvement limited to the four stages, namely (Rosemann 2018, pp. 176–179):

- navigate (identification of the current state)—the focus of the first stage is on the initial characteristics of the NESTT. This includes activities such as: deriving the process vision and essential attributes characterizing this vision, collecting and grouping ideas regarding possible future states for the three next stages, modeling the current value chain and a more detailed process using BPM notation, capturing physical and emotional states, KPIs, efforts and issues along the process (from the perspectives of customers and employees) and gathering information regarding resources (systems, people, documents), policies, and procedures;
- expand (vision of a future state)—this stage is dedicated to rapidly broadening the design space and deriving a comprehensive set of ideas with regard to: enhancing the existing process using improvement patterns such as elimination, resequencing, integration, or specialization, reactively generating ideas based on addressing issues, deriving ideas from different industries either in the form of general industry standards or by learning from specific organizations, utilizing idle assets which could include systems (big), data, employees, customers, physical assets, etc., and designing new experiences based on design-led innovation techniques;
- strengthen (project stage)—the third stage starts with the allocation of idea champions to each of the identified ideas. Based on individual expertise, passion and closeness to the required data and users, each member of the NESTT team will assume ownership of one or more ideas. Depending on the process and the data available, this will require making certain assumptions. This work is similar to the development of a brief business case (for each idea). In addition to the

analysis-intensive work of writing these mini business cases down, user validation is be required. Here, in line with design thinking principles, selected ideas are presented by the idea champions to and discussed with user groups allowing early feedback and important input to the further development of each idea. Finally—and this requires a dedicated session with a representative from the internal risk management team—each idea needs to be assessed in terms of risk and, where required, risk mitigation strategies need to be described. It is to be expected that after the validation, a number of initial ideas will be excluded from further consideration;

- tune/take off (implementation stage)—the fourth and final stage in the NESTT is all about ideas and getting these ready for the 'take-off'. The stage starts with developing a framework consolidating and interrelating all ideas, i.e. an investigation into any cause-effect relationships among the same. This could, for example, mean aligning the ideas with the value chain. This framework will also be used to calculate the total impact of the NESTT project on this process. Tuning every idea involves activities such as developing revised policies and procedures, crafting new forms as required by the new process or based on detailed assessment of the idea's compliance with external requirements. Job descriptions might need to be revised, which requires HR involvement, while IT changes will need to be discussed with the IT department. The most important milestone of the entire NESTT is presentation to the panel, i.e. a group of senior stakeholders who judge respective ideas and either endorse or reject the same. The decisions need to be documented and provide the go-ahead for the idea implementation, i.e. the take-off. A separate implementation team will work on accelerated implementation of the ideas. In many cases the same might entail an incubation phase during which an idea is further tested in one area (e.g. one organizational entity) before organization-wide roll out (take-off) of the (revised) idea.

Documented experiences with the NESTT have demonstrated that rigorous process change can be done quickly and that conducting such changes can be a highly valuable experience for everyone involved. In this regard, the three essential lessons learnt from the NESTT are in the areas of governance, participation (people), and facilitation (methodology) (Rosemann 2018, pp. 183–184):

- governance—it is essential to embed the rapid NESTT approach into an equally rapid governance structure. Otherwise, the NESTT loses its momentum and the desired accelerated idea-to-implementation is impossible to achieve,
- participants—the NESTT relies heavily on the creativity, energy, mindset, competence and attitude of the participants. Participants need to be receptive to the guidance of the facilitator. In particular, it is crucial to channel conversations into the right sessions, i.e. to decouple, for example, conversations regarding the current state from those considering weaknesses and possible solutions,
- methodology—most BPM methods and techniques are people-agnostic and ignore the impact of the facilitator on the quality of the outcomes. The NESTT is the opposite and the role of the facilitator is probably the most critical success factor. NESTT facilitators are not expected to be domain experts, but they must have the following characteristics: strong design capabilities and ability to develop future process scenarios, strong conceptualization and systemic thinking skills, being decisive and able to guide conversations in a limited time-frame, being able to work with stakeholders who are diverse in terms of seniority and attitude towards change.

6.2.4 Solidifying Process Improvements

Effective implementation of improvements is only the first step towards success, it is equally important to ensure that the achieved changes are measured and related to the previous performance of maturity level determined under the adopted model. This measurement of change (improvement or perfection) ought to be conducted with a number of key evaluation criteria in mind (Krogstie 2016, p. 78):

- accuracy—will be useful in measuring how well or to what extent the goal has been reached,

- objectivity—is important to ensure that the same conclusion will be reached regardless of who conducts the evaluation,
- goal dimensions—the main advantage of using more than one goal dimension is that it allows one to evaluate the results against different criteria of success. If the measures focus on just one dimension, one runs the risk of sub-optimization outside the same,
- specific target—a specified target will facilitate better evaluation of the result (e.g. a general target like "We want the process execution to be faster" is not as powerful as "We want the process execution to be 35% faster"),
- optimization—balancing the trade-offs between cost/quality and speed/flexibility, because lower cost will often entail decreased quality, while increasing speed may hinder flexibility—the aim is always to find a rational solution,
- goals and measures clearly communicated to everyone involved—to achieve an effective and productive working process, everyone has to pull in the same direction—understandable and motivational goals and measures are a prerequisite for this,
- compatibility with the organization's strategies—if some of the goals and measures are contrary to the organization's strategies, one will never be able to reach the main goal. As demonstrated in the context of psychology and organizational theory alike, goals that are set too high or too low too low can be equally demotivating.

By following the above guidelines, an organization can ensure that the implemented changes will be lasting but also viable (i.e. rational and justified) relative to overall goals of the organization (rather than only in the narrow context of the process itself).

6.3 Comparison and Recommendations—Anna Kosieradzka and Katarzyna Rostek

Process perfection is another stage during which differences between production companies and public administration institutions become apparent.

6.3.1 Organizational Context

The organizational context is of paramount importance in this case as it dictates the specifics of optimization criteria and directions of improvement that will be characteristic of the given subject matter. In production organizations, process perfection is subordinated to the implementation of a chosen approach in organizational management and, specifically, pertains to many processes and areas within the governance domain. When introducing changes in respective processes, one should therefore account for their influence and interactions observed at the tangency points with other processes.

In turn, in public administration institutions, the organization itself determines the directions and scope of improvement, but it should nonetheless do so with due consideration for the entire organization's role in its social and economic milieu. In accordance with the Lean Government approach, the generation of public value must take precedence—to an extent and at the level consistent with the applicable legal and social requirements. Consequently, every improvement effort ought to be preceded by an organization-wide analysis, e.g. with the use of the management quality self-assessment model.

6.3.2 Classical Approach to Improvement

Whenever attempting process perfection, it is important to assume a comprehensive approach covering not only the given process or fragment thereof, but also its interactions in the context of the entire organization and its other processes. As evidenced in numerous studies and practical applications, pointwise improvement of optimization of isolated process elements rarely proves effective in the global context. It is therefore important to strive towards improvement of processes considered in an organization-wide perspective, rather than only isolated fragments of the same.

It is clear that when drawing upon the experiences of production companies, public administration institutions must modify the introduced improvement measures to adapt them to their own functional specificity (Lean Management vs. Lean Government). Nonetheless, one has to emphasize the value of tested models that can be derived from the experiences of other entities, be it competitors or entirely different types of organizations, provided that the same prove to be at a higher level of process management maturity, given that they are duly adapted and dedicated to individual needs.

6.3.3 Contemporary Approach to Improvement

Both in production companies and public administration institutions, contemporary concepts of organizational improvement aim to meet the demand for fast and flexible methods facilitating efficient implementation of process changes. This acceleration can be observed in many dimensions and pertains not only to changes in terms of a given process, but also handling new challenges and requirements stemming from the emergence of new digital technologies that affect customers' expectations regarding the channels, means, and scopes of service availability, as well as the ever greater changeability of the economic, legal, social, and environmental milieu. All the above factors boil down to the conclusion that improvement ought to take the form of a continuous, cyclic effort, rather than a reactive measure, and that its methods should be selected with all the above mentioned requirements in mind.

Production companies perfect their processes by employing dedicated management methodologies (e.g. Lean Management, Kaizen, Six Sigma) which despite being rooted in the classical theory of management (e.g. Total Quality Management) successfully satisfy the requirements of flexibility, speed, and efficiency as well as other conditions indicated above. In turn, public administration tends to tackle these challenges by employing the agile concept which yields similar effects, but through a proper approach to the management of projects and undertakings.

6.3.4 Solidifying Improvements

There are two aspects of key importance to successful process improvement: 1) adopting a comprehensive perspective of the process in the context of the entire organization, and 2) being able to shift the focus of improvement from sporadic, reactive activities to continuous and global cycles. The latter condition is related to solidifying the effects of improvement but also reliable verification of its actual value and significance. Therefore, in long-term perspective, permanence the implemented changes should be evaluated through measures that verify their effectiveness, efficiency, and sufficiency relative to the initially adopted objectives. In production processes, regardless of the adopted methodology, such measures are well established and generally accepted standards. Public administration, on the other hand, has yet to consider the same more broadly and devote the time and effort necessary to define them. In both cases, however, effective measures are necessary if one aims to ensure that the success of a given implementation does not last only as long as the process or task is continued, but rather that its effects continue to be observable and measurable in the context of the organization's operations.

References

Aguirre, S., Parra, C., & Alvarado, J. (2012, June). Combination of Process Mining and Simulation Techniques for Business Process Redesign: A Methodological Approach. In *International Symposium on Data-Driven Process Discovery and Analysis* (pp. 24–43). Berlin, Heidelberg: Springer.

Bogan, C. E., & English, M. J. (2006). *Benchmarking jako klucz do najlepszych praktyk*. Wydawnictwo Helion.

Gordon M. (2010) *Six Sigma Quality for Business & Manufacture*. Palm Har,

Griffin, R. W. (2017). *Management*. Boston: Cengage Learning.

Hammer, M., & Champy, J. (1993). *Reengineering the Corporation: A Manifesto for Business Revolution*. New York: HarperCollins.

Heizer, J., & Render, B. (2011). *Operations Management* (10th ed.). Upper Saddle River, NJ: Pearson.

Imai, M. (1986). *Kaizen. The Key to Japan's Competetive Success*. New York: McGraw-Hill.

Jackiewicz, I., & Jackiewicz, A. M. (2017). Benchmarking: teoria i praktyka. *Przedsiębiorczość i Zarządzanie, 18*(2, cz. 2 Zarządzanie humanistyczne i publiczne), 117–131.

Kanawaty, G. (1992). *Introduction to Work Study*. Geneva: ILO.

Kosieradzka, A., & Ciechańska, O. (2018). Impact of Enterprise Maturity on the Implementation of Six Sigma Concept. *MPER, 3,* 59–70. https://doi.org/10.21008/j.0239-9415.2017.074.01.

Krogstie, J. (2016). Quality of Business Process Models. In M. Glykas (Ed.), *Quality in Business Process Modeling* (pp. 53–102). Cham: Springer.

Lunau, S. (Ed.). (2008). *Six Sigma + Lean Toolset*. Berlin, Heidelberg: Springer-Verlag.

Nadler, G. (1967). *Work System Design the Ideal's Concept*. Homewood: Irwin.

Oakland, J. S. (2014). *Total Quality Management and Operational Excellence*. New York: Routledge.

Oakland, J. S., & Oakland, R. J. (2019). *Statistical Process Control*. London and New York: Routledge.

Rosemann, M. (2018). The NESTT: Rapid Process Redesign at Queensland University of Technology. In J. Mendling & J. vom Brocke (Eds.), *Business Process Management Cases* (pp. 169–185). Cham: Springer.

Schwab, K. (2016). *The Fourth Industrial Revolution.* Cologny, Switzerland: World Economic Forum.

Shankar, R. (2009). *Process Improvement Using Six Sigma—A DMAIC Guide.* Milwaukee: ASQ Quality Press.

Sujova, A., Simanova, L., & Marcinekova, K. (2019). Reengineering of production processes and its impact on the financial situation and business performance of the company. *Engineering Management in Production and Services, Engineering Management in Production and Services, 11*(3), 106–116. https://doi.org/10.2478/emj-2019-0025.

7

Models of Process Maturity in Organizations

Anna Kosieradzka

7.1 Maturity Models—General Characteristics and Systematization—Anna Kosieradzka

The concept of maturity models emerged in response to the need for a method of measuring progress made by organizations in the process of continuous improvement. They aim to quantitatively express a variety of aspects related to an organization's operations. The underlying understanding of maturity in most studies devoted to this topic describe it as continuous development of the organization's competences in the given area, in an effort to improve performance.

A maturity model comprises a set of various tools and practices that facilitate the evaluation of an organization's management competence (OGC 2010) and serve to improve on key factors conditioning the achievement of adopted goals (Looy 2014). As defined by OMG (Object Management Group), maturity levels are evolutionary processes

A. Kosieradzka and K. Rostek, *Process Management and Organizational Process Maturity*,
https://doi.org/10.1007/978-3-030-66800-6_7

of implementing key practices in one or several areas of an enterprise's operations.

Attempts at defining models facilitating multidimensional evaluations of organizations have been in a variety of management domains. The greatest advances in the development of maturity models have been made in the areas of quality, project management, software development and production, as well as continuity management.

Table 7.1 presents the systematization of selected, most prevalent maturity models (Kosieradzka and Smagowicz 2016). In this classification, models of process management maturity were included as a separate group. They are general models that facilitate a comprehensive evaluation of processes in any given organization. One should note, however that more specific models focusing on respective areas of organizational operations have also been developed against the background of the process approach.

One of the most widely known models of process management maturity is the Business Process Maturity Model (BPMM) developed by the international Object Management Group (OMG) organization. It comprises thirty areas of evaluations (ascribed to the five respective levels of maturity) as well as nearly 350 particular practices (OMG 2008). In turn, the Business Process Maturity Model (BPMM) developed by Gartner consulting company identifies six areas of evaluation, with a set of guidelines developed in each area to facilitate classification under the adequate maturity level in a five-tier scale.

Another popular model in the context of process management is known as the Enterprise Maturity Model (EMM); it was proposed by Michael Hammer (2007), who identified: 5 process enablers (enablers) and 4 company capabilities: "*Together, the enablers and the capabilities provide an effective way for companies to plan and evaluate process-based transformation*". These process enablers determine how well a process is able to function over time. Companies that are able to put these enablers in place possess important enterprise-wide capabilities: their senior executives support a focus on processes; their employees greatly value customers, teamwork, and personal accountability; they employ people who know how to redesign processes; and they are well organized to tackle complex projects.

More specific models were identified in the areas of:

Table 7.1 Systematization of the most popular maturity models

Process management BPMM—OMG; BPMM—Gartner; PEMM)				Any organization
Software development management (CMMI, COBIT, ITSMMM)	**Quality management** (QMMG, ISO 9004, EFQM)	**Project management** PMMM—Kerzner, OPM3, P2MM, P3M3)	**Risk and continuity management** (BCMM, ERMMM, RMM-RIMS)	*Business organizations*
Production management (PMM, MTMM)				
Administration management (PRI)	**Quality management** (ISO 9004, CAF)		**Risk and continuity management** (RMPCM)	*Administration*
Operations management	Quality management	Project management	Risk and continuity management	

Source Based on Kosieradzka and Smagowicz (2016)

- Quality management
- Operations management
- Project management
- Risk management and continuity management

The first maturity models were proposed in the context of quality management, parallel to the development of the Total Quality Management concept. In 1979, Philip Crosby published his Quality Management Maturity Grid (QMMG), which is considered to be the first maturity model ever developed. It introduced a 5-point maturity scale that continues to be widely used in contemporary models. Evaluation is performed on a grid where the columns correspond to respective maturity levels and the lines specify six areas subject to assessment (Crosby 1980, pp. 21–34). In the late 1980s, ISO standards were published in the 9000 series which included the premise of a maturity model based on the ISO 9004 standard. It contains guidelines to the improvement of quality management systems, with the premise that the same will lead to better performance across the organization. Here to, a 5-point maturity scale was adopted. Notably, the latter system also introduced an organizational self-assessment tool that has since been used in many subsequent proposals. Another model commonly employed in the context of management is the EFQM Excellence Model developed by the European Foundation for Quality Management; it facilitates self-assessment, audits, and reviews in candidate organizations considered for the EFQM Excellence Award. The evaluation relies on nine criteria grouped into "enablers" (5 criteria) and "results" (4 criteria), and a 5-point maturity scale (0%, 25%, 50%, 75%, 100%) (Oakland 2014, pp. 156–168). The EFQM model is used primarily by business organizations (production companies and service providers), but a variant dedicated to public administration was also developed based on this model: CAF—The Common Assessment Framework.

Over the last decade, dynamic development of maturity models was also observed in the area of project management, both in the context of business entities and administration institutions. The first proposal in this context was the Kerzner Project Management Maturity Model—PMMM (Kerzner 2017, pp. 733–737) developed based on many years of practical experience. Currently, the area of project management seems to be dominated by the influence of two major organizations: the British Office of Government Commerce (OGC) and the American Project Management Institute (PMI). The former developed the PRINCE2 Maturity Model (P2MM) to evaluate organizations in terms of effective application of the PRINCE2 methodology, while in response to the growing demand in terms of programs and portfolios, the Cabinet Office (formerly OGC) developed the P3M3 model (P3O 2008). The model's structure is composed of three closely correlated "internal" models: portfolio management, program management, and project management—evaluation is performed in each of the areas independently, relative to seven knowledge aspects. In the 1990s, the latter of the mentioned organizations, Project Management Institute, developed its own maturity model called OPM3. Its structure comprises four elements: Best Practices, Capabilities, Outcomes, Key Performance Indicators, while the assessment itself is strongly correlated with the percentage share of respective practices employed by the organization (PMI 2003).

One of the first and simultaneously best-known maturity models dedicated to the processes of operations management was developed in the early 1990s by American programmers from the Software Engineering Institute (SEI) and was utilized for the purposes of improving software development processes. CMMI—Capability Maturity Model Integration was based on the Capability Maturity Model (CMM) and identified 4 areas: Process Management, Project Management, Engineering, and Support. It contained a total of 25 Process Areas associated with lists of do-called best practices without which the respective levels of maturity, on a 5-point scale, cannot be reached. The standard structure of the CMMI model comprising areas, goals, and related practices, along with its 5-point maturity scale, served as the foundation on which later models were developed.

In the context of IT management, the IT Service Management Maturity Model (ITSM Maturity Model) was developed by CTPartners S.A. The model's structure is composed of six domains with which between 3 and 8 specific processes were associated (https://itsm.zone/itsm-maturity/). Model Control Objectives for Information and related Technology (COBIT) were proposed by ISACA and IT Governance Institute with a view to gathering good practices in the area of IT process management. Evaluation is conducted within five identified domains and with respect to 37 processes (COBIT 5.0 revision), on a 6-point maturity scale. (https://www.isaca.org/resources/cobit).

Literature provides little information on maturity levels dedicated to production. A paper by Kasse (2008, pp. 429–437) includes a short reference to the implementation of CMMI in a production company, Israel Aerospace Industries Ltd. Also, the model described by Kobayashi (1995) might be considered a production maturity model based on the evaluation of performance in 20 areas related to production management. One could also mention the Industry 4.0 related maturity model named MTMM—ManuTech Maturity Model (Gracel and Łebkowski 2018). Kosieradzka (2017) proposed the Productivity Management Model developed on the basis of the CMMI, whose goal is to support production companies in the implementation of management concepts aimed at improving productivity.

In the context of Polish public administration, a maturity model was developed based on the Institutional Development Planning (IDP) method. The assessment here is based on a 5-point scale and 19 management criteria grouped into five key areas (www.pri.msap.pl).

In the 21st century, organizations are learning to appreciate the operating significance of forecasting possible adverse events as well as the negative effects thereof, with a view to implement adequate safeguards. The most popular model in the area of risk and continuity management is the Business Continuity Maturity Model (BCMM) developed by Virtual Corporation. The evaluation is based on a 6-point maturity scale, with 11 areas subject to assessment—enterprise competences (7 areas) and continuity criteria (4 areas) (Virtual Corporation Inc. 2020). Another model utilized in the area of risk and continuity management is the Enterprise Risk Management Maturity Model (ERM Maturity Model), which recognizes 8 areas of assessment analyzed on a 5-point maturity scale (Ciorciani and Blattner 2008). Based on the best-known maturity model—CMMI, the Risk and Insurance Management Society Maturity Model (RIMS-RMM) was developed in the context of risk management. It employs a 5-point scale and a survey conducted with respect to seven attributes, 25 competence directions, 68 key indicators that jointly characterize an organization. The advantage of tis models stems from the inclusion of numerous standards and norms applicable to risk and continuity management (www.rims.org). In terms of crisis management in the public sector, a model named RMPCM—Risk Management in Public Crisis Management was proposed (Smagowicz 2019).

Kosieradzka and Smagowicz (2016) analyzed 20 maturity models in seven distinct areas of management, including: process management, production management, project management, software development management, administration management, quality management, and risk and continuity management. The presented models were analyzed in terms of four aspects: structure, method of assessment, correlations with other concepts and methods, and implementation. The results of this study are summarized in Table 7.2.

Based on the presented characteristics, one can observe significant similarities between the models in terms of their structure and identified maturity levels. This is likely due to the fact that some models are developed based on earlier proposals (mainly CMMI) as well as that models are often adopted from one management domain to another; for instance the CAGF models was created by adapting the EFQM approach, which applies primarily to business entities, to the specific conditions of public sector institutions.

Most maturity models include a number key and closely correlated elements, such as (Kosieradzka and Smagowicz 2018):

- maturity level—degree of a given organization's maturity determined relative to its competences and skills,
- process areas—a set of activities whose effective performance in the given organization allows it to achieve its goals, as well as the diagnosed level of maturity,
- goals—indicating values ascribed to the respective process areas, whose achievement will allow the organization to reach a given maturity area,
- repository of good practices—collection of methods and techniques related to the respective process areas, whose successful implementation in the context of the organization's activities will allow it to

Table 7.2 Similarities between the analyzed maturity models

Analyzed aspect	Compared element	Similarity factor	Number of models including the factor
Structure	Number of maturity levels	4–6 levels	18
	Model structure	Areas and practices	18
	Number of areas	Up to 20 items	12
	Number of practices	Up to 500 items	10
Assessment method	Model representation	Continuous	4
		Stage	10
		Continuous and stage	6
	Scope of model application	Identification of the status quo	20
		Identification of the target status	20
	method of gathering data	Self-assessment	9
		Survey	6
		Observation based on adopted criteria	8
	Method of measuring results	Number of model criteria met	14
		Scoring system	6
	Person conducting the assessment (from within or outside the organization)	Person from outside the organization	16
Correlations with other models/concepts/business	Concepts employed	Continuous improvement	16

Analyzed aspect	Compared element	Similarity factor	Number of models including the factor
		Process approach	8
		Benchmarking	12
	Based on other models/systems	CMM	8
		Other models apart form CMMI	2
		Not applicable	10
	Certificates/awards won	Applicable	4
	Size of the enterprise	All	14
	Industry	All	14
Implementation	Ease of model implementation	High	13
	Time needed for implementation	Up to 6 months	14
	Effect of implementation	Organizational improvement	18
		No return on investment	17
	Implementation cost relative to effects gained	Low	11

Source Kosieradzka and Smagowicz (2016)

achieve goals adopted in respective process areas—and thus reach a specific maturity level.

The above elements are very closely correlated, as illustrated in Fig. 7.1.

The maturity of an organization is determined by verifying the satisfaction of requirements associated with the respective process areas in conjunction with the degree to which practices associated with a given area have been implemented and rated on the adopted scale. Maturity models use discrete scaling, usually within the range from 1 to 5, where "1" represents the lowest and "5" the highest level. The maturity scales adopted in selected models are presented in Table 7.3.

The evaluation of an organization's maturity with the use of maturity models can be conducted using stage or continuous representation. In stage representation, each maturity level is associated with a number of process areas (corresponding to the respective areas of the organization's activity) whose implementation allows the organization to reach a specific maturity level. The principle of cumulation applies, which means that the achievement of a given maturity level is possible only once all the process areas assigned to that level have been positively assessed, hence the improvement process follows a strict schedule. In turn, under continuous representation the maturity assessment is performed individually

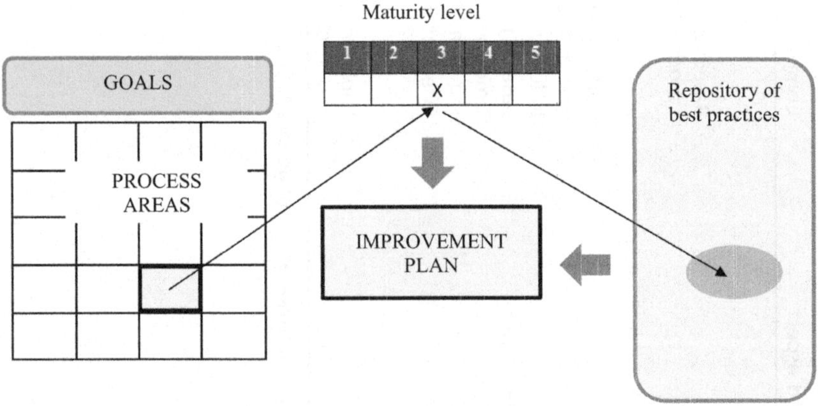

Fig. 7.1 General structure of maturity models (*Source* Kosieradzka and Smagowicz [2018])

Table 7.3 Maturity scales in selected maturity models

BPMM-OMG	CMMI	ISO 9004	QMMG (Crosby)
1. Initial 2. Managed 3. Standardized 4. Predictable 5. Innovating	1. Performed 2. Managed 3. Defined 4. Quantitatively Managed 5. Optimizing	1. No formal approach 2. Reactive approach 3. Formal systemic approach 4. Emphasis on continuous development 5. Best-in-class achievement	1. Uncertainty 2. Awakening 3. Enlightenment 4. Wisdom 5. Certainty
OPM3	BCMM	Kerzner	CAF
1. Standardize 2. Measure 3. Control 4. Continuously Improve	1. Self-Governed 2. Departmental 3. Corporative 4. Standards Compliance 5. Integrated 6. Synergistic	1. Common Language 2. Common Processes 3. Singular Methodology 4. Benchmarking 5. Continuous Improvement	0–10—Not active 11–30—Planning 31–50—Implementing 51–70—Checking 71–90—Adjust 91–100—PDCA (continuous improvement)

Source Own elaboration

for each process areas in accordance with the adopted maturity scale, which allows for considerable flexibility in the organization's choice of improvement activities. The key structural differences between maturity models developed in accordance with the stage or continuous representation approach are presented in Fig. 7.2.

The adopted maturity levels guide organizations in the development and improvement of their practices, starting from undefined and inconsistent practices and processes, through repeatable practices at the level of respective organizational units, followed by comprehensively defined business processes (predictable and statistically managed), all the way to continuous optimization and implementation of innovations. Maturity models are therefore a way of establishing the current condition of the organization based on its functional efficiency, ability to utilize available

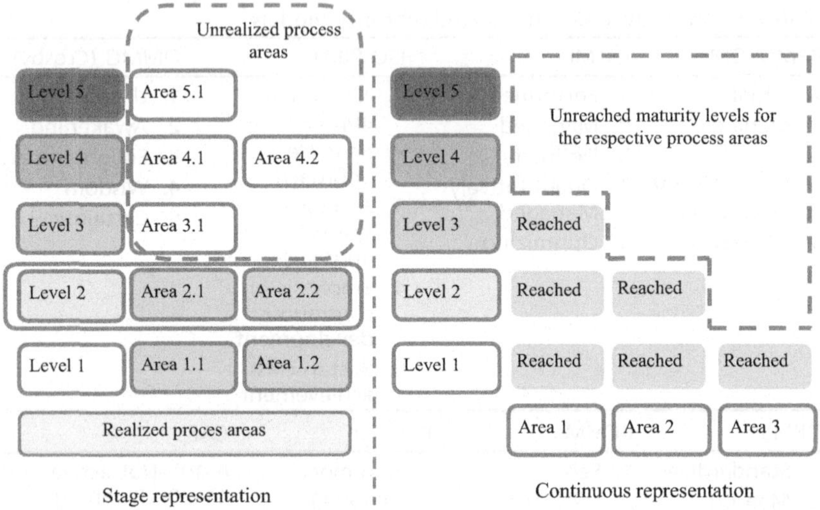

Fig. 7.2 Stage and continuous representation

resources or take advantage of previous experiences, but also determining goals that can be viably accomplished in the future by prioritizing, identifying resources, and directions of activity.

7.2 Process Maturity Models in Production Companies—Anna Kosieradzka

The following section will present examples of process maturity models dedicated to production organizations, notably the EFQM Excellence Model used in thousands of businesses, popular mainly in Europe but similar to its counterparts prominent in the US (Baldridge Criteria for Performance Excellence) or Japan (Deming Prize). The authors' own Productivity Management Model dedicated to production companies will also be presented as a proposal consistent with contemporary concepts of production management aimed at increasing productivity.

7.2.1 The EFQM Excellence Model

The EFQM Excellence Model is widely recognized in Europe as a tool facilitating comprehensive evaluation of production companies and service providers. It is a proprietary model developed by the organizer of the European Quality Award competition—the European Foundation for Quality Management (EFQM®), whose mission is to stimulate and support efforts contributing to lasting quality improvements. The EFQM model is rooted in the concepts of TQM and continuous improvement, it subscribes to the opinion that isolated and incidental activities aimed at functional improvements may prove ineffective. The EFQM Excellence Model (Fig. 7.3) considers nine criteria.

Five of the same can be classified as elements of an organization's potential (enablers) while the remaining four are described as results. The enabler criteria pertain to activities undertaken by the organization, while the results criteria consider its accomplishments. The results are products of the enablers, while the enablers are improved based on feedback received by measuring the results. Excellent results in terms of activity, customers, people, and society are attainable under leadership that serves

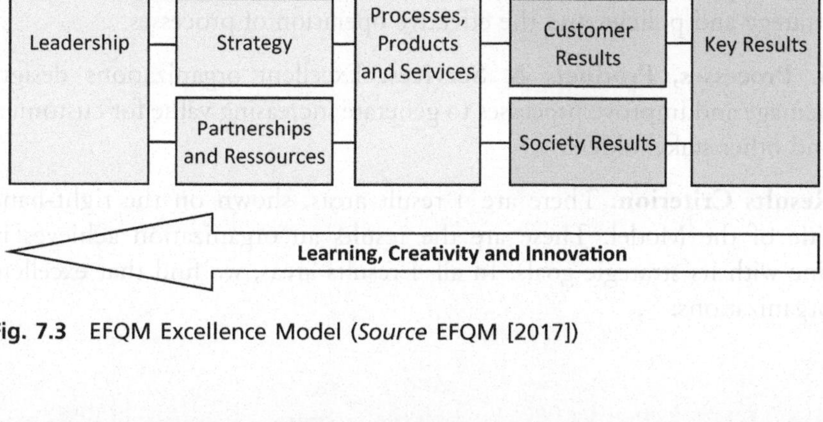

Fig. 7.3 EFQM Excellence Model (*Source* EFQM [2017])

as the driving force for policies and strategies implemented through the contributions of people, partnerships, resources, and processes. The arrows are used to emphasize the dynamic character of the Model. They indicate that innovations and learning help to increase the organization's potential, which ultimately leads to better results.

Each of the nine criteria is provided with a definition that explains its significance and identifies the applicable sub-criteria (EFQM 2017).

1. Leadership. Excellent organizations have leaders who shape the future and make it happen, acting as role models for its values and ethics and inspiring trust at all times. They are flexible, enabling the organization to anticipate and react in a timely manner to ensure the on-going success of the organization.

2. Strategy. Excellent organizations implement their mission and vision by developing and deploying a stakeholder focused strategy. Policies, plans, objectives and processes are developed and deployed to deliver the strategy.

3. People. Excellent organizations value their people and create a culture that allows the mutually beneficial achievement of organizational and personal goals. They develop the capabilities of their people and promote fairness and equality. They care for, communicate, reward and recognize, in a way that motivates people, builds commitment and enables them to use their skills and knowledge for the benefit of the organization.

4. Partnerships & Resources. Excellent organizations plan and manage external partnerships, suppliers and internal resources in order to support strategy and policies and the effective operation of processes.

5. Processes, Products & Services. Excellent organizations design, manage and improve processes to generate increasing value for customers and other stakeholders.

Results Criterion. There are 4 result areas, shown on the right-hand side of the Model. These are the results an organization achieves in line with its strategic goals. In all 4 results areas, we find that excellent organizations:

- Develop a set of key performance indicators and related outcomes to determine the successful deployment of their strategy, based on the needs and expectations of the relevant stakeholder groups.
- Set clear targets for key results, based on the needs and expectations of their business stakeholders, in line with their chosen strategy,
- Segment results to understand the performance of specific areas of the organization and the experience, needs and expectations of their stakeholders,
- Demonstrate positive or sustained good business results over at least 3 years,
- Clearly understand the underlying reasons and drivers of observed trends and the impact these results will have on other performance indicators and related outcomes,
- Have confidence in their future performance and results based on their understanding of the cause and effect relationships established,
- Understand how their key results compare to similar organizations and use this data, where relevant, for target setting.

6. Customer Results. Excellent organizations achieve and sustain outstanding results that meet or exceed the needs and expectations of their customers.

7. People Results. Excellent organizations achieve and sustain outstanding results that meet or exceed the needs and expectations of their people.

8. Society Results. Excellent organizations achieve and sustain outstanding results that meet or exceed the needs and expectations of relevant stakeholders within society.

9. Business Results. Excellent organizations achieve and sustain outstanding results that meet or exceed the needs and expectations of their business stakeholders.

The complete maturity model comprises 32 sub-criteria representing the detailed requirements of the 9 main criteria. The weight of the respective criteria and sub-criteria is expressed as in percent. The percentage prescribed for a given area reflects that areas significance in the entire self-assessment model, whereas the weight of sub-criteria—their relative contribution to the evaluation of the given area. By analyzing all functional aspects of an enterprise with the use of a scoring system, one is able to determine the level of the entire organization's development and excellence. The EFQM model can be used as a self-assessment tool useful in determining the distance covered by the organization on its way towards excellence as well as understanding its shortcomings and finding solutions.

The self-assessment questionnaire includes questions pertaining to all the respective criteria and sub-criteria. Responses constitute evaluations given on a scale from 0 to 100. The information is subsequently aggregated based on the percentage weights adopted for the respective sub-criteria to obtain values corresponding to the nine main criteria (thus producing the so-called score profile of the organization presented in the form of a radar graph) (Fig. 7.4).

Each of the concentric axes corresponds to one criterion of the EFQM model. The sum of the scores constitutes the synthetic organization-wide result with the maximum value of 1,000.

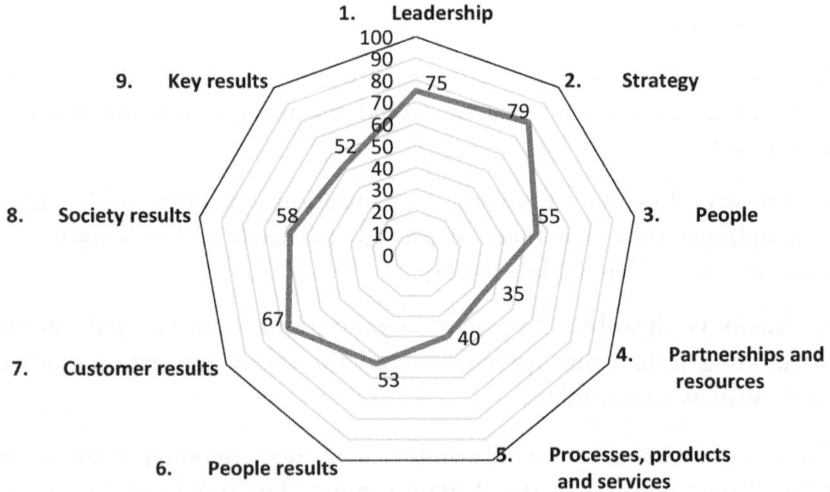

Fig. 7.4 Sample score profile of an organization (*Source* Own elaboration)

The EFQM self-assessment survey should be conducted as follows:

- The questionnaire should be completed independently by several people (5–10), preferably members of the given organization's top management,
- The obtained scores should be averaged,
- He results should be presented in a form of pointwise profile for each of the areas,
- The organization's strengths should be identified as well as areas in need of improvement,
- The organization's overall score should be calculated.

The results of the self-assessment allow the identification of the organization's strengths and areas that require improvement. Strengths are the areas where the obtained result is the highest while areas in need of improvement are the ones that were scored the lowest. Notably, the Model does not consider the latter weaknesses but rather areas to be improved.

Self-assessment is the basis for the subsequent planning of improvement measures that should focus on the areas that:

- received the lowest score,
- are the most important in terms of the organization's strategy and priorities adopted by the management.

Improvement efforts ought to be based on so-called good management practices formulated based on the output of management sciences. When planning improvement activities, it is advisable to take advantage of the good practices register available from the national partner organization of the EFQM.

In the subsequent versions of the model developed since 2010, the primary emphasis of the rules and criteria has been on sustainable organizational development. This means making an effort to viably balance:

- the needs of customers and employees,

- quality and costs,
- innovativeness and costs,
- environmental concerns and economic growth,
- building a market presence and generating profits.

The key benefits of employing the EFQM model include:

- insight into the organization as a whole (comprehensive assessment),
- identification of strengths and areas in need of improvement,
- evaluation of the effects of introduced improvements (optimization of intra-organizational processes),
- increased employee involvement in the undertaken activities (building a pro-quality organizational culture),
- propagation of the ideas of learning and creativity,
- monitoring and control of the direction of organizational growth,
- ability to measure progress in time by conducting periodic self-assessment,
- ability to compare the organization's score with other similar organizations provided they are evaluated under the same model),
- sustainable growth.

7.2.2 PMM—Productivity Management Model

Continuous improvement in production management is aimed at increasing enterprise productivity understood as the relationship between the outputs (goods and services) generated from a production system and the inputs (labor, tangible and financial capital assets, materials, energy, information) needed to create such outputs (Kosieradzka 2012, p. 19).

Nowadays, the following concepts aimed at increasing productivity are commonly used in production management: LM—Lean Manufacturing, TOC—Theory of Constraints, TQM—Total Quality Management, SS—Six Sigma, TPM—Total Productive Maintenance, Kaizen, BPR—Business Process Reengineering, and 5S (Kosieradzka 2012, p. 52). These concepts are usually implemented independently of each other or are only incidentally linked. Consequently, the interactions and tools shared between the same are not sufficiently utilized. At the same time, the concepts are so closely related that it is sometimes hard to tell whether a given solution is implemented as part of TQM or LM. Is TPM an independent system employed in production management or—as some say—a part of LM? In recent years, we have observed increasing popularity of a concept referred to as Lean Six Sigma—a system combining LM and Six Sigma (Lunau 2008, pp. 5–10), as well as of so-called TLS—a combination of TOC, LM and Six Sigma (Cox and Schleier 2010, pp. 1067–1080). Moreover, many methods and techniques are shared between different concepts. For example, 5S practices and Kaizen philosophy can be traced present in all concepts, SMED is considered a tool of LM and TPM, and SPC (Statistical Process Control) is a key element in TQM, Six Sigma, and Lean Manufacturing.

Treating existing management concepts separately leads to diverse organizational structures, documentation systems, training programs etc. The lack of coordination makes enterprise management very complicated and means that the potential for synergy, stemming from the fact that a lot of methods and specific techniques are common to different models, is squandered. At the same time, uncoordinated implementation of different concepts often leads to failure as organizations attempt to use highly advanced tools without first taking advantage of less complex methods. It is therefore clear that pursuing organizational excellence in the field of production management requires simultaneous and coordinated implementation of all the aforementioned management concepts.

The above considerations encouraged the development of the productivity management model presented in (Kosieradzka 2017). Its aim is to arrange the processes of implementing contemporary concepts of production and productivity management with a view to effectively improving enterprise productivity.

The main purpose of the Productivity Management Model is to support enterprises in the implementation of modern management concepts geared towards productivity improvement by means of:

- Identifying the current status of implementation;
- Identifying the desired (target) status;

- Identifying methods and techniques that must be implemented to transition from the current to the target status.

The general structure of the Productivity Management Model is shown in Fig. 7.5.

The Productivity Management Model may be treated as a production process maturity model in an enterprise. Its starting point is the implementation of 5S practices at the level of individual workstations, as a clean, orderly, organized, and productive workplace is necessary to facilitate the subsequent introduction of comprehensive systems in the areas of production management (LM and TOC), quality management (TQM and Six Sigma), and technical resources maintenance (TPM). The process approach lies at the core of all those systems. The focus on processes allows gradual identification, modeling, and improvement of processes on the path to their eventual optimization, and in specific cases even to radical process reengineering in line with the BPR methodology.

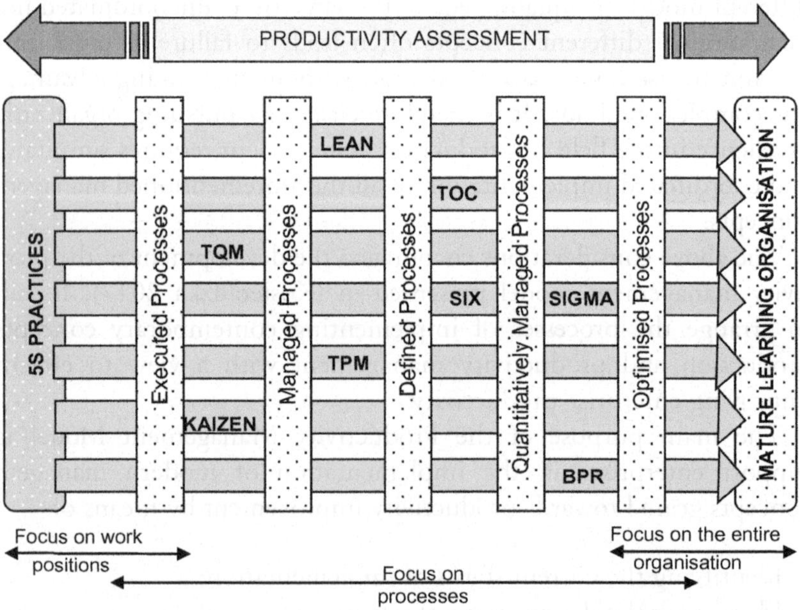

Fig. 7.5 Productivity Management Model (*Source* [Kosieradzka 2017])

Kaizen serves as the philosophy underlying the implementation of LM, TOC, TPM, TQM, and Six Sigma systems with its emphasis on continuous improvement of all processes in an enterprise by relying on knowledge, experience, and engagement of the entire workforce. People constitute a key resource in this context; hence the model assumes a continuous focus on the employees and deployment of adequate HR management instruments, in particular with a view to incentivization and improvement of qualifications. Indeed, the changes affecting human resources are so thorough that they result in the development of an entirely new organizational culture, one that could be referred to as a "culture of productivity".

The implementation of the described methods is to allow continuous productivity improvement whose progress should be measured using productivity indicators (see Chapter 5.1). Ultimately, this is to gradually transform the company into a learning organization, i.e. one that "*will focus on continuous improvement of management processes, goods and services, support people development and continuous change for improved execution of the strategy and stronger commitment of all employees to ensure better adaptability to the volatile global environment*" (Senge 2006, pp. 3–5).

The respective components of the Productivity Management Model were defined based on the CMMI model and include:

- The Concept: a comprehensive approach to solving problems in production management, involving the design of new and application of existing methods and techniques. The concept uses a range of methods and techniques designed to achieve its primary objective (e.g. Lean Manufacturing). The model builds on the following concepts: 5S, LM, TOC, TQM, Six Sigma, TPM, Kaizen and BPR.
- The Area: a delineated subset of methods and techniques within the concept which, if implemented together, will facilitate meeting the objectives that aggregate to the overall goal of the concept (e.g. rapid changeover is a LM component). Specific management concepts were selected for the Productivity Management Model form one or several areas (for more complex concepts). PMM identifies a total of 17 Areas.

Each Area is scored on a scale of 1 to 5 and individual area scores yield the total enterprise maturity score.

- Practices: methods, techniques, measures, procedures, and tools assigned to respective Areas, the application of which helps to meet the Area Objectives. Each Concept has a dedicated toolbox of methods and techniques designed to facilitate the achievement of its Goal and Objectives.
- Maturity Level: a term derived from CMMI denoting how advanced an organization is in organizing and managing its production processes. For the purpose of the model at hand, the levels are called production maturity levels. Specific levels are outlined in Table 7.4.

Table 7.4 Maturity levels for production management

Level 1	Effectively executed production processes	Targets are met (adequate products are manufactured in the right quantity and on time)—but these processes are not iterative or predictable, which makes it impossible to control progress
Level 2	Managed production processes	Production targets are met as a result of implementing plans and progress is monitored for consistency with plans
Level 3	Defined production processes	Production targets are met in processes defined (described) in line with the process approach parameters
Level 4	Quantitatively managed production processes	Quantitative and qualitative targets and performance control tools have been defined for individual processes and their constituents (operations)
Level 5	Optimized production processes	Processes are continuously improved and adapted to the changing environment and corporate strategy

Source Based on Chrissis et al. (2003, pp. 75–82)

- The current production maturity profile represents levels currently accomplished by the organization in each area.
- The target production maturity profile represents a set of areas and the associated levels expected after the implementation of certain practices the organization intends to adopt within a certain time horizon.
- Reference maturity profile is enterprise-specific, i.e. it depends on the industry, the size of the organization, type of manufacturing, corporate culture etc. It specifies the benchmarked achievable levels for respective areas. The reference level represents world-class manufacturing level. It helps in guiding the long-term strategy of productivity improvement in the organization.

In the Productivity Management Model, consecutive production maturity levels are achieved by implementing adequate practices assigned to specific areas and designed to help accomplish the goals and objectives. The model supports the identification of the current production maturity profile and setting the target profile. The gap between the target profile and the current profile guides the planning of synergistic productivity improvement measures (implementing general practices applicable to several areas) to ensure sustainable growth of the business (with a focus on even growth in all areas).

The current production maturity profile can be assessed using an assessment questionnaire designed for this model. It helps in identifying the exact level on the scale to reflect the current status of the organization. The scale is designed to ensure that levels achieved in specific areas correspond to the production maturity level of the entire organization. Table 7.5 illustrates the current and target production maturity profile for a sample organization (Kosieradzka 2010).

The PMM model was tested in a study group of several dozen production enterprises sharing the following characteristics (Kosieradzka 2017):

- Medium or large enterprise;
- At least several years in business;
- Representing different manufacturing types (job, batch, mass);

Table 7.5 Maturity profile of a sample enterprise

Concept	Area	Maturity level				
		1	2	3	4	5
5S	5S Practices					
LM	Waste elimination					
	LM principles					
	Leveling – one-piece flow					
	VSM - Value Stream Mapping					
	Pull system – kanban					
	Standardization					
	Supplier partnering (JIT)					
TPM	SMED					
	TPM - principles					
TQM	Mistake proofing – Poka Yoke					
	Quality management system					
	TQM - principles					
Six Sigma	Six Sigma - principles					
Kaizen	Kaizen – Continuous improvement					
TOC	TOC - principles					
BPR	Process management and BPR					

Legend: (light grey) Target production maturity profile; (dark grey) Current production maturity profile

Source Kosieradzka (2017)

- Representing several different industries (e.g. electromechanical engineering, food-processing, automotive, industrial automation and household appliances);
- Forming homogenous subgroups of several entities for which a joint reference profile can be developed.

The results of the study revealed that the Productivity Management Model may prove helpful to manufacturing organizations aiming for productivity improvement in the area of production management. This conclusion was corroborated by all participating respondents. The model promotes a balanced and coordinated implementation of productivity improvement concepts, methods and techniques in line with the idea of sustainable organization development, not unlike the EFQM Excellence Model.

7.3 Models of Process Maturity in Public Administration

The following section will discuss examples of process maturity models dedicated to public administration. The deliberations will be based on our own as well as secondary studies.

7.3.1 Organizational Self-Assessment Under the CAF Model—Katarzyna Rostek

One of the tools most commonly used and recognized in public administration for the purposes of multidimensional organization evaluation is the CAF model—Common Assessment Framework (Fig. 7.6). It allows entities to conduct self-assessment in the process of continuous improvement, both at the scale of a single organization and more broadly, e.g. in benchmarking studies.

CAF is a total quality management tool developed by the public sector for the public sector, inspired by the Excellence Model of the European Foundation for Quality Management (EFQM). This nine-box structure identifies the main aspects requiring consideration in any organizational analysis. Criteria 1–5 (Enablers) deal with the managerial practices of an organization. These determine what the organization does and how it approaches its tasks to achieve the desired results. In criteria 6–9 (Results) achieved in the fields of citizens/customers, people, social responsibility and key performance are measured by perception

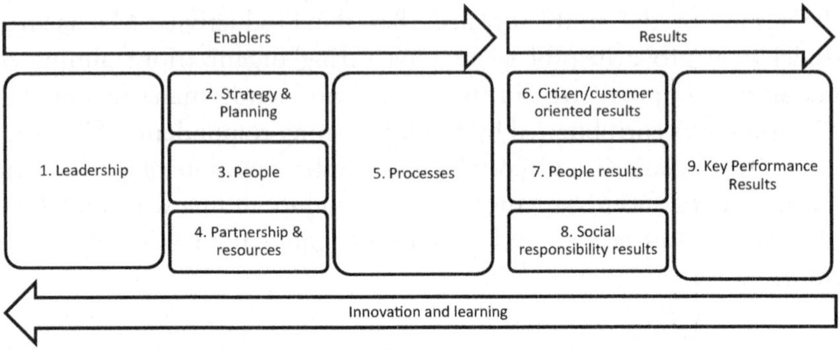

Fig. 7.6 Common Assessment Framework Model (*Source* CAF-2020, p. 11)

and performance measurements. The respective criteria can be described as (CAF 2020, pp. 14–52):

- Leadership—the leaders of the organization must clearly and precisely declare its vision, mission, and objectives. They should communicate this to the personnel and demonstrate personal engagement in the related efforts and continuous improvement of the organization. Moreover, it is the task of the top management to motivate and support employees in the performance of their tasks, with a particular focus on maintaining as good relations with customers and other stakeholders as possible,
- Strategy & Planning—the task of the organization is to develop strategies and operating plans based on reliable criteria, e.g. customer or employee surveys. Strategies ought to be periodically reviewed and updated and the results of their implementation should be evaluated by e.g. benchmarking against other similar organizations,
- People—the priority is to ensure the introduction of adequate procedures to facilitate proper management of the personnel, e.g. in terms of training, education support, initiating dialogue between employees and the management, encouraging teamwork, and introducing clear and flexible work rules,
- Partnership & Resources—the organization must aim to maintain good relations with external partners. It should also adopt solutions that will allow it to optimize the management of its resources

(especially financial), information flows (within and without the organization), as well as IT solutions,

- Processes—every organization, regardless of its location, size, or structure, should identify its processes, properly manage and improve the same to ensure that the goals, strategies, and plans it adopts are fully compatible with the needs and expectations of its external and internal customers,
- Citizen/customers Oriented Results—the organization is expected to continuously analyze and meet the needs of customers, but also to increase their satisfaction with the scope, manner, and quality of the services rendered,
- People Results—the organization ought to analyze and make efforts to increase the satisfaction of its personnel. This serves the purpose of retaining and developing highly qualified and competent staff members capable of performing their assigned tasks at the expected level of quality,
- Social Responsibility Results—the social responsibility of public administration institutions is particularly high given their obligation to create public value for the society. Hence, the organization ought to study and aim to satisfy the local community, as well as minimize its negative impact on the natural environment,
- Key Performance Results—the organization should measure, evaluate and continuously improve the effects of its financial and non-financial activities.

Each general criterion is further broken down into a list of sub-criteria. The 28 sub-criteria identify the main issues that need to be considered when assessing an organization. The analysis considers the organization simultaneously from different angles—it is a holistic approach to a performance analysis based on eight key principles (CAF 2020, p. 8):

- results orientation—the organization focuses on results. Results are achieved which please all of the organization's stakeholders (authorities, citizens/customers, partners and people working in the organization) with respect to the targets and responsibilities that have been set,

- citizen/customer focus—the organization focuses on the needs of both current and potential citizens/customers. It involves them in the development of products and services and the improvement of its performance. This is to increase awareness of applicable needs and encourage greater involvement of customers in the creation of services and products,
- leadership and constancy of purpose—this principle couples visionary and inspirational leadership with constancy of purpose in a changing environment. Leaders establish a clear mission statement, as well as a vision and values, they also create and maintain the internal environment in which people can become fully involved in realizing the organization's objectives,
- management by processes and facts—this principle guides the organization from the perspective that a desired result is achieved more efficiently when related resources and activities are managed as a process and effective decisions are based on the analysis of data and information,
- people development and involvement—people at all levels are the essence of an organization and their full involvement enables their abilities to be used for the organization's benefit. The contribution of employees should be maximized through their development and involvement and the creation of a working environment of shared values and a culture of trust, openness, empowerment and recognition. It should be remembered that a satisfied employee will carry out the duties entrusted to him with much greater care and precision, and the client's welfare will be a priority for him,
- continuous learning, innovation and improvement—excellence is challenging the status quo and effecting change by continuous learning to create innovation and improvement opportunities. Continuous improvement should therefore be a permanent objective of the organization. However, it is important that this goal is implemented in a planned manner and based on real and rational premises,
- partnership development—public sector organizations very often need others to achieve their targets and should therefore develop and maintain value-adding partnerships. An organization and its suppliers are

interdependent, and a mutually beneficial relationship enhances the ability of both to create value,
- social responsibility—public sector organizations have to assume their social responsibility, respect ecological sustainability and try to meet the major expectations and requirements of the local and global community.

These principles are integrated into the structure of the CAF Model and the continuous improvement of the nine criteria in accordance with the defined procedure (Fig. 7.7) will in time bring the organization to a high level of maturity.

The value of the self-assessment performed is largely determined by the method of assigning individual assessments. It should be remembered that allocating a score to each sub-criterion and criterion of the CAF model has four main aims (CAF 2020, p. 53):

- to provide information and give an indication of the direction and priorities to follow for improvement activities,
- to measure progress regularly—once every two years is considered to be good practice according to most quality approaches,
- to identify good practices as indicated by high scores recorded for enablers and results,

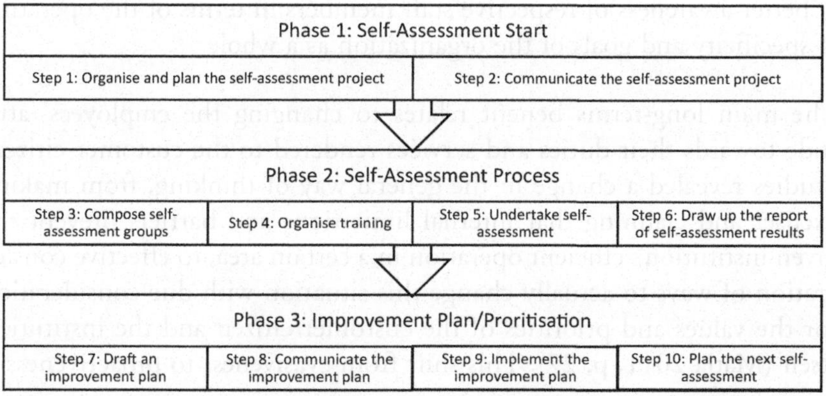

Fig. 7.7 Common Assessment Framework Procedure (*Source* CAF [2020, p. 60])

- to help to find valid partners to learn from—benchlearning (what we learn from each other).

Criteria measured according to the above principles using classical (CAF 2020, pp. 53–54) or fine-tuned (CAF 2020, pp. 55–59) scoring methods identify priorities and areas for improvement and prioritize them. The results obtained at the stage of process verification coupled with data obtained from the above analysis provide a source of comprehensive knowledge needed to ensure the viability, purposefulness, and rationality of the improvement stage.

The analyzed public administration organizations indicated short- and long-term benefits stemming from regular application of the CAF method. In terms of short-term benefits, the respondent institutions identified (Matla 2011, p. 26):

- the ability to monitor and conduct multidimensional assessment of the institution's operations,
- quality improvements in various functional areas of the institution (after improvement based on the results of CAF assessment),
- identification of the institution's strengths and weaknesses,
- increased employee engagement and participation,
- improvement of the organizational awareness, qualifications, and skills of employees,
- better awareness of respective staff members in terms of the operating specificity and goals of the organization as a whole.

The main long-terms benefit relates to changing the employees' attitude towards their duties and services rendered to the customer-citizen. Studies revealed a change in the general way of thinking, from making excuses and claiming that internal limitations and barriers prevent the given institution's efficient operation in a certain area, to effective consideration of ways to actually change this situation with due consideration for the values and priorities of the customer/citizen and the institution itself (Matla 2011, p. 27). This shift from evasiveness to proactiveness is the main reason why correct application of CAF becomes a viable investment facilitating the organization's growth and improvement, not only in terms of process management but management in general.

7.3.2 RMPCM—Risk Management in Public Crisis Management—Anna Kosieradzka

In the context of public crisis management, the primary social expectation towards the state relates to the assurance of national security understood as protection of citizens against various threats and their consequences. Public administration institutions are obliged to prepare crisis management plans. Even at the lowest administrative level, territorial self-government authorities must be ready to analyze and assess risks and prepare adequate strategies for managing crises, including methods of preventing threats and response should the same materialize.

The level of local administrative districts' preparation for the performance of tasks related to risk and crisis management can vary significantly. Efforts aimed at country-wide standardization of risk management systems can be facilitated by a maturity model serving as the basis for the assessment of local competences in this area and identification of the most advantageous directions for development in this respect. The RMPCM model (risk management in public crisis management) was developed in the course of studies conducted at the Faculty of Management of the Warsaw University of Technology (Smagowicz 2019). The model's structure is based on the general premise of the CMMI model and comprises four primary components: process areas, goals—closely related to the process areas, repository of good practices, and maturity levels.

7.3.2.1 Process Areas and Goals

The process areas of a maturity model determine the scope of requirements applicable to organizations in terms of risk management in crisis management. The thus identified requirements are closely related to the implementation of specific actions whose success is expected to yield improvement in the given area, and consequently improvement of the associated risk management process (Smagowicz 2019, p. 84).

The process areas of the model were described with reference to the subsequent stages of the risk management process in the context of crisis management:

- Context of the process—focusing on the identification and description of the object of protection in terms of: purpose of the conducted risk analysis and method of its realization, description of threat vulnerability (depending on the geographical location of administrative organization), or information on the entity conducting the analysis,
- Identification of threats—i.e. identification of all phenomena or events that may potentially pose a danger to the analyzed organization—the activity is aimed at compiling a list of threats to which the analyzed organization is susceptible, as well as any correlations between the respective threats,
- Risk analysis—focused on diagnosing the underlying causes, mechanisms, and potential consequences of risks—the activity is aimed at developing criteria for the assessment of likelihood and potential consequences of risks,
- Risk assessment—assessment of the likelihood and potential consequences of each of the identified risks (in terms of: casualties/injuries, material damage, or environmental impact)—the activity is to yield quantified risks and a complete risk matrix,
- Risk handling—focused on developing procedures for activities related to risk tolerance, prevention, and monitoring—the activity is aimed at developing risk handling plans, including in terms of continuity management.

Additionally, three universal areas present in many of the analyzed maturity models were included in the system, namely: leadership, partnership, and employee relations.

The correlations between the risk management process (five process areas) and overall management processes (3 process areas) are presented in Fig. 7.8.

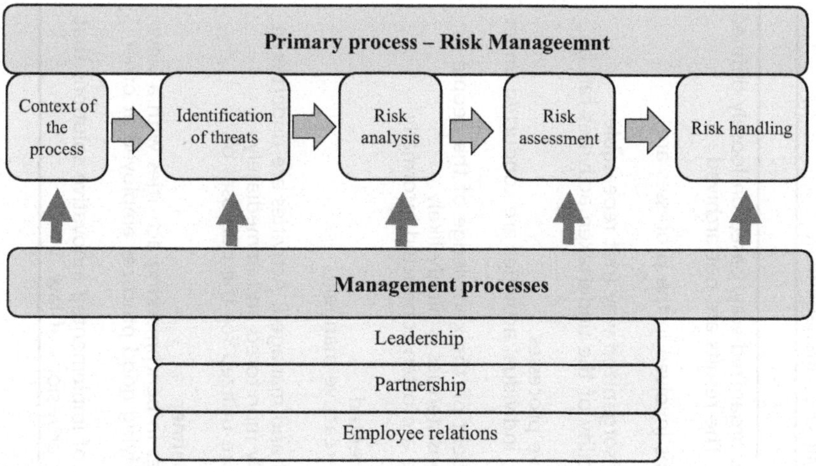

Fig. 7.8 Correlations between the primary processes and management processes constituting process areas in the original maturity model (*Source* Smagowicz [2019, p. 103])

Each area was related to specific goals that condition the satisfaction of requirements applicable to the given process area. In turn, the goals closely reflect requirements pertinent to risk management in the context of public crisis management or specific measures taken in this respect. The defined goals are accomplished by way of implementing adequate practices identified and systematized in the repository of good practices (Smagowicz 2019, p. 85).

7.3.2.2 Maturity Levels

The maturity levels of the RMPCM model delineate the scope of competences and skills demonstrated by the given organization in the context of the specific process area. The maturity assessment follows the method of continuous representation, i.e. each identified process area is evaluated individually on a five-point scale (from 1—the lowest to 5—the highest level). The general requirements applying to organizations at the respective levels of maturity are presented in Table 7.6

Table 7.6 Maturity levels in the maturity model for risk management in public crisis management

Maturity level	Description of requirements
Level 1	Processes realized at the institution are performed in a disorganized way, based on loosely defined guidelines, which means their results are unpredictable. The results are not archived Activities are undertaken reactively The roles and scopes of responsibility of respective persons involved in the processes are underdefined
Level 2	Processes realized at the institution are performed in a disorganized way (not repeatable, unpredictable). Nonetheless, some instances of repeatability of the undertaken activities can be observed despite the lack of formal guidelines Measurable goals have not been defined for the respective processes Despite high similarity between the actual tasks involved, individual activities are conducted under separate procedures The performance of respective processes is strictly dependent on the knowledge of the people involved, consequently the emergence of gaps and inconsistencies is highly likely Employees participate in necessary trainings but the same are often conducted informally
Level 3	The institution accomplishes most of its adopted goals Processes are well defined and described in detail, documented Most activities are undertaken in a proactive rather than reactive manner Reports and reviews are conducted regularly
Level 4	The processes realized at the institution are fully defined and managed. Activities are undertaken based on the adopted organizational goals, continuously monitored and immediately reported—the results indicated in reports and reviews are utilized for the purposes of improvement planning Activities are undertaken in a predominantly proactive manner The public administration institution continuously engages in benchmarking activities with a view to assessing its functional performance as well as identifying good practices employed at other institutions Cyclical evaluations are performed to analyze the results of implemented innovative solutions that are not compulsory under the organization's policies or provisions of law

Maturity level	Description of requirements
Level 5	Results of benchmarking analyses are positive and indicate that the institution achieves best-in-class results. Good practices developed by business and public administration organizations alike are continuously monitored and implemented as applicable Processes realized by the organization are adapted to evolving circumstances with significant contributions from the management The ability to improve processes stems from organizational procedures and is not dependent on the competences of respective process participants The public administration institution engages in the process of continuous improvement as a necessary condition of maintaining its competitive advantage Advances achieved by the institution in the area of crisis management are presented and emphasized in periodic internal and external communication

Source Smagowicz (2019, p. 125)

7.3.2.3 Repository of Good Practices

The repository of good practices is a collection of various tools assigned to the respective identified process areas, whose effective implementation in the organization facilitates the satisfaction of requirements applicable within those areas. From the perspective of management, the definition of a good practice is broad enough to include known and employed methods (ways of solving problems), various documents and procedures (describing guidelines, premises), as well as rules that should be followed in the context of a given activity—hence the scope of tools included in the repository of good practices is also rather extensive (Kosieradzka and Smagowicz 2018). Good practices have been identified for and associated with all eight of the relevant process areas. Such practices ought to include instruments that can be applicable to many areas of the maturity assessment (general practices), as well as methods and techniques dedicated to only one process area (specific practices). The distinction between general and specific practices was present in the CMMI model as well as other models developed on its basis (Kosieradzka and Smagowicz 2018).

7.3.2.4 Maturity Assessment

Maturity assessment is performed by determining the scope of requirements/practices implemented in the organization—which serves as the basis for the identification of the current maturity level, followed by a recommendation of the directions for organizational improvement complete with a schedule for the implementation of "lacking" requirements/practices in a specific sequence. The model-based maturity assessment is done in two perspectives: internal evaluation (self-assessment) performed by an employee, and external evaluation performed by an independent expert. The results are presented in a special evaluation sheet illustrated in Table 7.7. The described activity allows one to gain a full insight into the organization's situation and identify the desirable directions of its development. The role of expert may be served by a member of another self-governmental institution's personnel (e.g.

Table 7.7 Results of a maturity assessment conducted for a sample local self-government authority

Process areas	Sub-areas	Maturity level				
		1	2	3	4	5
Context of the process	Identification of the object of protection					
	Identification of entity conducting the assessment					
	Detailed description of the object of protection					
	Final assessment					
Identification of threats	Review and selection of threat identification methods					
	Identification of threats					
	Classification of threats					
	Identification of correlations (domino effect)					
	Final assessment					
Risk analysis	Specification of the consequence criterion					
	Specification of the likelihood criterion					
	Final assessment					
Risk assessment	Risk valuation					
	Specification of risk level					
	Final assessment					
Risk handling	Tolerance					
	Prevention					
	Monitoring					
	Continuity planning					
	Final assessment					
Leadership						
Partnership						
Employee relations						

Reached maturity level (green)

Target maturity level (red)

Source Smagowicz (2019, p. 208)

form a different region) with adequate experience in the performance of maturity analyses.

conducting the maturity assessment with the use of the presented model is relatively straightforward for anyone with access to sufficient information regarding activities undertaken by the organization in terms of risk management in crisis management. The result of the assessment will reflect the current state of affairs in the organization as well as indicate recommendations for future goals (in the context of organizational improvement).

The development of concepts pertinent to public administration management (new public governance and participatory public governance) led to changes in the ways that states are governed but also contributed to the implementation of new improvement strategies in public administration. Institutions started to look at solutions already successfully deployed in business organizations with a view to adopting the same in their own decision-making systems. One example of such solutions are maturity models whose systematized structure allows one to derive a quantifiable result (description of five maturity levels) from inherently qualitative analyses (description of activities undertaken by a given organization). Moreover, maturity assessment facilitates systematization of tools used in the organization and knowledge derived therefrom in the form of the repository of good practices. In turn, the repository may be used both in the ongoing assessment of a public administration institution's activity, and for the purposes of benchmarking between different self-governmental entities.

7.4 Comparison and Recommendations—Anna Kosieradzka

The purpose of this section is to summarize the above analysis of two distinct approaches to maturity assessment – respectively applicable to economic and non-economic organizations, reiterate the differences and similarities between the same, and use this information to present viable recommendations for practitioners, managers, and decisionmakers.

The above analysis of maturity models used in business entities and public institutions allowed the identification of models utilized in various areas of management. On the one hand, public administration organizations have a narrower range of models at their disposal in the context of continuous improvement processes, but on the other, the existing models available to economic entities tend to show a lot of conceptual and structural similarities.

Structurally speaking, most models include between four and six maturity levels, with five being by far the most common number. The structures of respective models are most often based on the identification of respective process areas to which general and specific goals are ascribed along with specific practices (methods and techniques) facilitating their achievement. The division into more homogenous process areas and definition of the related goals greatly facilitates subsequent analyses and evaluations. It has been observed that in most of the models, the number of identified areas and practices tended to be, respectively, under 20 and under 500. Greater numbers of such elements are primarily found in models referring to the domain of software development management due to more complex nature of the processes involved and more extensive correlations between the same. Only in some models was a lower number of areas accompanied by more than 500 items included in the repository of good practices. As for the method of conducting the assessment, most models opt for continuous representation that allows a more comprehensive overview of the organization. However, a growing number of models, particularly in the context of software development management, are now leaning towards simultaneous inclusion of continuous and stage representation.

All the discusses maturity models are characterized by broad applicability, both in terms of diagnosing the current status of an organization and determining the target status to be reached through improvement. The authors of the models took advantage of the continuous improvement concept and facilitated organizations in answering the question of "where does the organization find itself currently?", but also did not shy away from indicating the directions for future development. In terms of the methods of gathering data, self-assessment was by far the most popular—a tool derived from quality management concepts. The advantage of self-assessment lies in its ability to reflect the perception of the organization by its own employees who have access to detailed information concerning its functional capacity, but also to increase the employees' awareness of a more general context in the course of the assessment. The most commonly used techniques include surveys and direct observation, while interviews are employed primarily in assessments conducted by external experts as they can provide information identifying not only the current state of affairs but also the underlying causes thereof.

Most models employ similar methods of measuring results by determining the level of criteria satisfaction for respective areas on a point or percentage scale. In situations where the relative significance of respective areas is different, a weighing system is employed.

Due to the significant reliance of maturity models on the technique of self-assessment, most models lean towards evaluations performed by a person from within the organization, which is understandable given the models' focus on continuous improvement. In quickly developing areas such as project management, software development management, or risk and continuity management, it is nonetheless recommended to simultaneously account for both dimensions: internal assessment—easier to perform and more focused, but at the same time more subjective, as well as external assessment—more difficult and performed more broadly by an outside expert. External assessment is a requirement when applying for certifications granted exclusively by authorized institutions whose experts are competent to conduct the relevant assessments.

Maturity models are related to various concepts enjoying widespread prominence in contemporary management. Of the latter, the process approach is particularly significant to all the discussed models. Other notable concepts include continuous improvement, as an immanent element of process management, and benchmarking, which facilitates improvement through comparison to other organizations and adoption of their most effective solutions, so-called good practices.

The analyzed systems also showed a clear tendency to heavily draw upon the CMMI model. The adopted elements are found primarily in the models' structures (5-point maturity scale, classification based on process areas and practices) as well as the assessment methodology.

In terms of the direct advantages of implementing maturity models, many authors indicate that there may be not measurable benefits. In several models pertaining mainly to the areas of quality management and software development, model implementation and the related efforts may help an organization in working towards a specific certification (e.g. standard compliance) or award (e.g. EFQM Awards). The effects of synergy are also increasingly evoked to indicate that a comprehensive approach to an organization's development, collaboration between various departments or functional areas, may ultimately yield better results than individual improvements limited to selected areas. The applicability of maturity models in companies of different sizes and representing different industries was also analyzed. By principle, such models are developed with a view to solving problems inherent in a given industry but can nonetheless be adapted to also facilitate other sectors. Indeed, it seems that most of the maturity models are universal enough to be deployable in virtually any organization, regardless of its size or domain.

The more complex a model is and the more elements it contains, the more difficult it becomes to perform the assessment, which usually means it will also be more time consuming and costly. There is therefore a tendency to minimize the structure and rely on self-assessment, which facilitates easier interpretation, shorter implementation timeframes, and lower costs relative to the achieved effects.

It has been suggested that most of the benefits resulting from improving an organization's maturity level do not directly translate to a return on the invested capital. Nonetheless, organizations still decide to engage in such endeavors due to less tangible benefits such as better chances to win tenders (likelihood of receiving better scores) or opportunities to start collaboration with new customers. One of the key effects resulting from the implementation of a maturity model relates to performance improvements in the organization, which are reflected by corresponding KPIs.

Maturity models facilitate a wide analysis of an organization and its comprehensive assessment in terms of key requirements (e.g. legal acts, management concepts adopted, or internal policies) and stakeholder expectations (customers, contractors, employees, communities, etc.).

Different maturity model are suitable to different areas of management as each of the latter is connected to different factors affecting the overall growth of the organization. It runs out, however, that despite certain disparities, such models also have quite a lot in common. This is mainly due to the fact that the purpose of all maturity levels is to facilitate continuous improvement and perfection in an organization—understood as continuous expansion and increasing utilization of the range of skills, competences, and knowledge available to the organization in order to cope with the emerging challenges by introducing so-called best practices developed in the given area. Every model utilizes a repository of best practices that jointly constitute general, universal practices found in multiple models, as well as specific practices dedicated to solving specific problems emerging under specific models.

The attractiveness of organizational maturity models stems from their attempt to systematize the multitude of areas and processes that contribute to said maturity, precisely define the criteria that have to be met in order to reach given levels in particular areas, as well as associate the respective levels with a set of methods and techniques (best practices) recommended in the given context. Domain-specific maturity models developed with various areas of management are primarily used as a means of:

- continuous improvement of organizations
- benchmarking its accomplishments—by comparing maturity profiles of different organizations
- stimulating sustainable growth of organizations.

In the macro scale, the concept of sustainable development, originally developed by economists, refers to forms of socioeconomic development wherein political, economic, and social endeavors are integrated with care for the environmental equilibrium and preservation of elementary natural processes in order to guarantee the possibility of satisfying the basic need of respective communities or individuals in the context of the present as well as future generations. In recent years, the notion of sustainable development has been adopted in various fields, including management, by assuming that the same refers to achieving a certain balance between the needs of customers, employees, and communities, between cost, quality and innovation, ecology and economic growth, etc. The core idea is to look for ways of connecting various areas of an organization which ensure that none of said areas are ignored and all are developed at a comparable pace to ensure a harmonious growth of the organization as a whole. Models of organizational maturity aim to facilitate such sustainable growth by imposing a certain order in the implementation of respective methods, techniques, and activities corresponding to the particular modeled areas.

It should be emphasized, however, that the development of dedicated maturity models in respective domains is a task that requires substantial expert knowledge and considerable effort. The practical implementation of such models also entails not inconsequential expenditures and great organizational efforts. Given the rather prevalent unwillingness of people to accept overly strict procedures, one can also expect some resistance to excessive formalization that may be imposed by such models. On the other hand, the fact that quality models, in particular the EFQM model underlying the European Quality Award, are now generally accepted suggests that also maturity models may soon be widely embraced by organizations seeking a competitive edge. Attempts are also likely to be made with the aim f integrating domain-specific models into a single, universal system with a view to utilizing synergy effects given the fact than many of the so-called best practices are found in more than one model, which means that their separate implementation may be somewhat inefficient.

References

CAF. (2020). *The Common Assessment Framework (CAF). The European Model for Improving Public Organisations Through Self-Assessment*. European Institute of Public Administration.

Chrissis, M. B., Konrad, M., & Shrum S. (2003). *CMMI® Guidelines for Process Integration and Product Improvement*. Upper Saddle River, NJ: Adison-Wesley.

Ciorciani M., & Blattner P. (2008). *Enterprise Risk Management Maturity-Level Assessment Tool*. https://www.soa.org/globalassets/assets/files/resources/essays-monographs/2008-erm-symposium/mono-2008-m-as08-1-ciorciari.pdf.

Cox, J. F., & Schleier, J. G. (Eds.). (2010). *Theory of Constraints Handbook*. New York: McGraw-Hill.

Crosby P. (1980). *Quality is Free. The Art of Making Quality Certain*. New York: McGraw-Hill.

EFQM. (2017). *Overview of the EFQM Excellence Model*. EFQM.

Gracel, J., & Łebkowski, P. (2018). The Concept of Industry 4.0 Related Manufacturing Technology Maturity Model (Manutech Maturity Model, MTMM). In *Decision Making in Manufacturing and Services, 12*, 17–31. AGH University of Science and Technology, Kraków. ISSN: 2300-7087. https://journals.agh.edu.pl/dmms/article/view/2779.

Hammer, M. (2007). The Process Audit. *Harvard Business Review, 85*(4), 111–123.

Kasse, T. (2008). *Practical Insight into CMMI®*. Norwood: Artech House.

Kerzner, H. (2017). *Project Management: A Systems Approach to Planning, Scheduling, and Controlling*. Hoboiken: Wiley.

Kobayashi, I. (1995). *20 Keys to Workplace Improvement*. Cambridge, MA: Productivity Press Inc.

Kosieradzka, A. (2010). Applying Capability Maturity Model CMMI® to Production Management. In M. Fertsch (Ed.), *Innovative and Intelligent Manufacturing Systems* (Monograph, Chapter 8, pp. 115–130). Poznań: Publishing House of Poznan University of Technology.

Kosieradzka, A. (2012). *Zarządzanie produktywnością w przedsiębiorstwie* [Productivity Management in the Entrprise]. Warszawa: C. H. Beck.

Kosieradzka, A., & Smagowicz J. (2016). Analiza porównawcza modeli dojrzałości organizacji [Comparative Analysis of Organization Maturity Models]. In M. Ćwiklicki, M. Jabłoński, & S. Mazur (Eds.), *Współczesne koncepcje zarzadzania publicznego* (pp. 283–296). Kraków: Fundacja Gospodarki i Administracji Publicznej.

Kosieradzka, A. (2017). Maturity Model for Production Management. In K. Halicka & L. Nazarko (Ed.), *Procedia Engineering* (Vol. 182, pp. 342–349). Special Issue 7th International Conference on Engineering, Project, and Production Management. Elsevier. https://doi.org/10.1016/j.proeng.2017.03.109.

Kosieradzka A., Smagowicz J. (2018) *Model dojrzałości organizacji w obszarze publicznego zarządzania kryzysowego* [Organisation Maturity Model in the Area of Public Crisis Management]. In *Zeszyty Naukowe Politechniki Śląskiej* (pp. 217–230). Seria: Organizacja i Zarządzanie z. 128.

Looy, A. (2014). *Business Process Maturity: A Comperative Study on a Sample of Business Process Maturity Model*. Berlin, Heidelberg: Springer International Publishing.

Lunau, S. (Ed.). (2008). *Six Sigma + Lean Toolset*. Berlin, Heidelberg: Springer-Verlag.

Matla, J. (2011). *Dobre praktyki zidentyfikowane podczas samooceny w ramach projektu systemowego*. Warszawa: Ministerstwo Administracji i Cyfryzacji.

Oakland, J. S. (2014). *Total Quality Management and Operational Excellence*. New York: Routledge.

OGC. (2010). *Management of Risk: Guidance for Practitioners*. London: TSO.

OMG. (2008). *Business Process Maturity Model (BPMM)*. https://www.omg.org/spec/BPMM/1.0/PDF.

P3O. (2008). *Portfolio, Programme and Project Offices*. Norwich: Office of Governmental Commerce, TSO.

PMI. (2003). *Organizational Project Management Maturity Model (OPM3)*. Newton Square: Project Management Institute.

Senge, P. (2006). *The Fifth Discipline*. New York: Currency Doubleday.

Smagowicz, J. (2019). Model dojrzałości zarządzania ryzykiem w publicznym zarządzaniu kryzysowym, rozprawa doktorska [Risk Management Maturity Model in Public Crisis Management] (Doctoral Dissertation). Warsaw University of Technology, Faculty of Management, Warsaw.

Virtual Corporation. (2020). *Business Continuity Maturity Model*. https://www.virtual-corp.com/bcmm.

Summation

We live in a world where the phenomena of Industry 4.0 and ubiquitous digital transformation are forcing organizations to ever more precisely define the processes underlying their activities and continuously improve their ability to effectively manage the same. Based on many years of our own studies and observations derived from secondary research, we arrived at the thesis that the success or failure of an attempt to implement the process approach is largely dependent on the level of its adaptation to the specific type and needs of a given organization. And should one consider various existing classifications, the most distinctly different groups thereof are those of economic and non-economic organizations. This conclusion inspired us to describe the various stages and process lifecycles and problems of process management maturity from the perspective of production companies (representing economic organizations) and public administration institutions (representing non-economic organizations). In this context and assuming the perspective of a process lifecycle, an attempt was made to answer two research questions.

A. Kosieradzka and K. Rostek, *Process Management and Organizational Process Maturity*,
https://doi.org/10.1007/978-3-030-66800-6

Q1: To what extent can one identify differences in the process approach applicable to economic and non-economic organizations?

At the stage of process modeling in production companies, we usually identify ordered and normalized activities whose arrangement and scope stems directly from the primary process and is then complemented by the inclusion of adequate supporting and management processes. Hence in this case, the stage of modeling can and indeed should be used to tract, analyze, and evaluate the logic, suitability, and rationality of such efforts and the adopted sequence thereof. In turn, public administration institutions which have, so far, successfully operated in line with the functional approach, perceive the process approach as a method of increasing functional efficiency and directly improving their social value. This is why each newly developed process model tends to be original and unique, but also the reason for the increased risk of such a model's inadequacy or mismatch relative to the actual needs of the organization if incorrect process approach is employed.

Regardless of the type of organization, the implementation of a process must begin with specification of its business justification and referring the same to the strategic goals of the organization. In the context of production companies, we are once again faced with well documented and ordered activities that have always been present, even if not fully recognized for what they were. The challenge here is related not so much to process implementation as such, but rather to ensuring that the introduced version of the process is optimized and perfectly adapted to the goals and resources of the organization. Conversely, in the case of public administration institutions one must face a project of new process implementation which should be handled in accordance with generally accepted project management methods. The choice of methodology and specific project approach is rather significant here as both should reflect the scope and complexity of the given implementation project.

The stage of process verification serves a control function. It answers questions related to the quality, efficiency, and productivity of process performance, but above all quantifies effectiveness relative to the degree of process goal achievement. Production companies that have relied on processes since their establishment, tend to have suitable and tweaked sets of measures and indicators on hand, which not only greatly facilitates control but also guarantees comparability and standardization of processes (necessary e.g. when implementing Industry 4.0 solutions). Meanwhile, public administration institution does not have predefined measurement packages at their disposal, but this provides them with the opportunity to more precisely adapt measures and indicators to the specific needs of respective organizations. In this context, it is important to be able to effectively dimension the conducted activity so that the developed measures provide a precise but also relatively easy to interpret description of the organization's operations. Any experience in the formulation of KPIs or other indicators of operating efficiency will be invaluable in this context.

Process perfection is discussed in the study in two distinct perspectives: (1) in terms of individual changes introduced to a single process, and (2) as an organization-wide cycle of process management improvement and development. In both cases the organizational context is of key importance as it dictates the viable directions of changes as well as their correlation with the organization's strategy and objectives. While in the case of production companies the focus is primarily on the product's value to the customer and the process of its manufacture as such, in the case of public administration the central concern is to generate greater public value, and therefore create a more dependable public image rooted in the high quality of provided public services. It can therefore be concluded that while the approaches applicable to the two cases are characteristic of the respective types of organizations, the overriding objective remains the same—to improve performance, be it in terms of achieving business goals (production companies) or fulfilling the organization's mission (public administration).

P2: What is the importance of accounting for said differences relative to the efficiency and effectiveness of process results?

When it comes to process management, all organizations utilize a similar set of tools and techniques, their adopted methods and approaches are often also fairly similar. What distinguishes them, however, is the effectiveness thereof. Based on the preliminary consideration of the problem one could conclude that the decisive factors in this respect, similarly to any other implementation of a complex project, include the precision and accuracy with which the external and internal factors are taken into account, as well as the ability to contextually adapt the tools and methods at the organization's disposal.

Precise identification of the specific internal and external circumstances relevant to an organization aiming to transition to the process approach or improve the effectiveness thereof, is undoubtedly a key success factor conditioning the viability of the desired results. Internal circumstances are understood here as those related to the type of the organization, its size, scope and complexity of operations, capacity and qualifications of the managerial staff, as well as level of technological advancement. All those elements are important when determining the applicable process map and its respective constituents. External conditions include relations with the immediate and more distant milieu, external stakeholders, as well as any formal, legal, political, social, economic, or environmental factors that may apply.

It is noteworthy, however, that the impact of external and internal factors on the organization's processes will vary depending on the type of organization as well as the type of a given process as such. Evidently, while for production companies the primary significance must be attributed to product quality and its compatibility with market and customer expectations, in the case of public administration, despite the undeniable importance of satisfying the needs of customers, one cannot deny that formal, legal, social, and political requirements are the elements that directly dictate the range of offered services as well as the manner and scope of their rendition. Hence, the type of the given organization must necessarily be taken into account if one is to understand its operations and adequately design its processes with a view to ensuring their maximum quality and efficiency.

Similarly, the choice of suitable tools and methods will largely depend on the type of the organization in question, and consequently its goals, operations, market outreach, size, and personnel qualifications, as well as the type and level of technological solutions employed. As demonstrated above, production companies, as originators of processes and the process approach as such, may and indeed should be treated as a viable benchmark for other types of organizations, including public institutions, that are only beginning to recognize the benefits and limitations inherent in process management. However, if such models are applied incompetently, without duly accounting for the type and specificity of the given organization, the same is bound to be ineffective and may even lead to grave errors of judgement. It seems that the adaptive approach may be the most recommendable here as it is able to "translate" known and popular management concepts and methodologies (e.g. Lean Management) into versions thereof adapted to the specific requirements of particular organization types (e.g. Lean Manufacturing, Lean Government). One can expect that in time and as new experiences are gained, we may eventually reach a point where packages of standard methods, tools, techniques, and approaches will be versioned and dedicated to respective organization types, thus ensuring the maximum efficiency and effectiveness of their implementation.

It is our hope that the monographic study succeeded in demonstrating the significance and viability of accounting for the type of the organization whose processes are to be modeled, implemented, controlled, or improved. Our aim was to inspire researchers and practitioners to make the effort of adaptation, and in doing so achieve better process performance by employing the best suited, comprehensive approach in their management practices.

Bibliography

Aguirre, S., Parra, C., & Alvarado, J. (2012, June). Combination of Process Mining and Simulation Techniques for Business Process Redesign: A Methodological Approach. In *International Symposium on Data-Driven Process Discovery and Analysis* (pp. 24–43). Berlin, Heidelberg: Springer.

Ahrend, N., Walser, K., & Leopold, H. (2012). Comparative Analysis of the Implementation of Business Process Management in Public Administration in Germany and Switzerland. In K. Sandkuhl, U. Seigerroth, & J. Stirna (Eds.), *Emerging Topics in the Practise of Enterprise Modeling* (pp. 1–12). 5th IFIP WG 8.1 Working Conference, PoEM 2012.

Andrews, R., Suriadi, S., Wynn, M., ter Hofstede, A. H., & Rothwell, S. (2018). Improving Patient Flows at St. Andrew's War Memorial Hospital's Emergency Department Through Process Mining. In *Business Process Management Cases* (pp. 311–333). Cham: Springer.

Batko, R. (2009). Urząd gminy jako organizacja procesowa. *Zarządzanie publiczne. Zeszyty Naukowe ISP UJ*, *4*(8), 53–64.

Bernus, P., Nemes, L., & Schmidt, G. (2012). *Handbook on Enterprise Architecture*. Berlin and Heidelberg: Springer Science & Business Media.

A. Kosieradzka and K. Rostek, *Process Management and Organizational Process Maturity*,
https://doi.org/10.1007/978-3-030-66800-6

Bitkowska, A. (2019). *Od klasycznego do zintegrowanego zarządzania procesowego w organizacjach* [From Classical to Integrated Process Management in Organizations—Only in Polish]. Warszawa: C.H. Beck.

Bogan, C. E., & English, M. J. (2006). *Benchmarking jako klucz do najlepszych praktyk*. Wydawnictwo Helion.

BPTrends Report. (2018). https://www.redhat.com/cms/managed-files/mi-bptrends-state-of-bpm—2018-survey-analyst-paper—201803-en.pdf. Read 22 July 2020.

Buijs, J. C., La Rosa, M., Reijers, H. A., van Dongen, B. F., & van der Aalst, W. M. (2012, June). Improving Business Process Models Using Observed Behavior. In *International Symposium on Data-Driven Process Discovery and Analysis* (pp. 44–59). Berlin, Heidelberg: Springer.

Cabinet of Ministers Regulation of October 29, 2008 on the Polish Classification of Goods and Services (PKWiU), Dz. U. [Journal of Laws] No. 207, item 1293, as amended.

CAF. (2020). *The Common Assessment Framework (CAF). The European Model for Improving Public Organisations Through Self-Assessment*. European Institute of Public Administration.

Capgemini. (2012). *Global Business Process Management Report*.

Cappelli, C., & Leite, J. C. S. P. (2010). Software Transparency. *Business and Information Systems Engineering, 2*, 127–139.

Cherouana, A., Mahdaoui, L., Bellatreche, L., & Medjahed, B. (2019). A Semantic Approach for Generating Government Processes. *International Journal of Web and Grid Services, 15*(1), 59–92.

Chrissis, M. B., Konrad, M., & Shrum, S. (2003). *CMMI® Guidelines for Process Integration and Product Improvement*. Upper Saddle River, NJ: Adison-Wesley.

Ciorciani, M., & Blattner, P. (2008) *Enterprise Risk Management Maturity-Level Assessment Tool*. https://www.soa.org/globalassets/assets/files/resources/essays-monographs/2008-erm-symposium/mono-2008-m-as08-1-ciorciari.pdf.

Cox, J. F., & Schleier, J. G. (Eds.). (2010). *Theory of Constraints Handbook*. New York: McGraw-Hill.

Crosby, P. (1980). *Quality Is Free: The Art of Making Quality Certain*. New York: McGraw-Hill.

Ćwiklicki, M., & Walczak, M. (2015). Modelowanie procesów jako podstawa standaryzacji kosztów w jednostkach samorządu terytorialnego. *Nauki o Zarządzaniu, 3*(24), 41–57.

Davenport, T. H. (1993). *Process Innovation: Reengineering Work Through Information Technology*. Boston: Harward Business School Press.

Devjak, S., & Peček, B. (2004). Simulation Techniques in Public Administration Procedures. *Economia Aziendale Online, 1*(2004), 1–17.

Dietrich, E., & Schulze, A. (1999). *Statistical Procedures for Machine and Process Qualification.* Milwaukee: ASQ Quality Press.

Dimian, A. C., Bildea, C. S., & Kiss, A. A. (2014). Introduction in Process Simulation. In *Computer Aided Chemical Engineering* (Vol. 35, pp. 35–71). Elsevier.

Dumas, M., La Rosa, M., Mendling, J., & Reijers, H. A. (2013). *Business Process Management*. Berlin: Springer-Verlag.

Dumas, M., La Rosa, M., Mendling, J., & Reijers, H. A. (2018). *Fundamentals of Business Process Management* (2nd ed.). Berlin: Springer.

EFQM. (2017). *Overview of the EFQM Excellence Model*. EFQM.

Engiel, P., Araujo, R., & Cappelli, C. (2014). Designing Public Service Process Models for Understandability. *Electronic Journal of e-Government, 12*(1), 95–111.

Feld, W. M. (2001). *Lean Manufacturing: Tools, Techniques, and How to Use Them.* Boca Raton: The St. Lucie Press.

Feldkamp, D., Hinkelmann, K., & Thönssen, B. (2007, October). Kiss–Knowledge-Intensive Service Support: An Approach for Agile Process Management. In *International Workshop on Rules and Rule Markup Languages for the Semantic Web* (pp. 25–38). Berlin, Heidelberg: Springer.

Ferlie, E. (2017). The New Public Management and Public Management Studies. *Oxford Research Encyclopedia of Business and Management*. https://doi.org/10.1093/acrefore/9780190224851.013.129

Goldratt, E. M. (1990). *The Haystack Syndrome: Sifting Information Out of the Data Ocean.* Croton-on-Hudson, NY: North River Press.

Goldratt, E., & Cox, J. (1992). *The Goal*. Great Barrington: The North River Press.

Gordon, M. (2010). *Six Sigma Quality for Business & Manufacture*. Amsterdam: Elsevier.

Gracel, J., & Łebkowski, P. (2018). The Concept of Industry 4.0 Related Manufacturing Technology Maturity Model (Manutech Maturity Model, MTMM). *Decision Making in Manufacturing and Services, 12,* 17–31. AGH University of Science and Technology, Kraków. ISSN: 2300-7087. https://journals.agh.edu.pl/dmms/article/view/2779.

Griffin, R. W. (2017). *Management*. Boston: Cengage Learning.

Grünwald, P. D. (2007). *The Minimum Description Length Principle.* Cambridge: MIT Press.

Hammer, M. (2007). The Process Audit. *Harvard Business Review, 85*(4), 111–123.

Hammer, M. (2010). *What Is Business Process Management?* In J. vom Brocke & M. Rosemann (Eds.), *Handbook on Business Process Management 1*. International Handbooks on Information Systems. Berlin, Heidelberg: Springer-Verlag.

Hammer, M., & Champy, J. (1993). *Reengineering the Corporation: A Manifesto for Business Revolution*. New York: HarperCollins.

Harrington, H. J. (1991). *Business Process Improvement: The Breakthrough Strategy for Total Quality, Productivity and Competetiveness*. New York: McGraw-Hill.

Harris, R., Harris, C., & Wilson, E. (2003). *Making Materials Flow*. Brookline: Lean Enterprise Institute.

Heizer, J., & Render, B. (2011). *Operations Management* (10th ed.). Upper Saddle River, NJ: Pearson.

Hevner, A. R., March, S. T., Park, J., & Ram, S. (2004). Design Science in Information Systems Research. *MIS quarterly*, pp. 75–105.

Hinkelmann, K., Probst, F., & Thönssen, B. (2006). Agile Process Management: Framework and Methodology. In *AAAI Spring Symposium: Semantic Web Meets eGovernment* (pp. 33–41).

Hood, C., & Jackson, M. (1992). The New Public Management: A Recipe for Disaster? *Hazard Management and Emergency Planning: Perspectives on Britain* (pp. 109–125). London: Routledge.

http://organizacjahoryzontalna.pl/procesowosc-to-sie-oplaca/. Read 18 October 2019.

http://www.bptrendsassociates.com/about-us/bpm-methodology/. Read 3 July 2020.

https://fr.boc-group.com/uploads/files/_2015__ADONIS_Whitepaper_-_EN_04.pdf. Read 13 July 2020.

https://itsm.zone/itsm-maturity/dostęp. Read 18 October 2020.

https://sceweb.uhcl.edu/helm/RationalUnifiedProcess/process/workflow/environm/co_imppr.htm. Read 29 July 2020.

https://www.ariscommunity.com/event-driven-process-chain. Read 13 July 2020.

https://www.isaca.org/resources/cobit/dostęp. Read 18 October 2020.

https://www.omg.org/spec/BPMN/2.0/PDF. Read 13 July 2020.

https://www.omg.org/spec/UML/2.5.1/PDF. Read 13 July 2020.

https://www.pri.msap.pl. Read 18 October 2020.

https://www.rhythmsystems.com/blog/how-to-implement-a-new-process-company-wide. Read 29 July 2020.

https://www.rims.org. Read 19 October 2020.

Imai, M. (1986). *Kaizen. The Key to Japan's Competetive Success.* New York: McGraw-Hill.

Imai, M. (1997a). *Gemba Kaizen: A Commonsense, Low-Cost Approach to Management*. New York: McGraw-Hill.

Imai, M. (1997b). *Kaizen: The Key to Japan's Competitive Success.* New York: McGraw-Hill.

ISO 9000:2015. Quality Management Systems—Fundamentals and Vocabulary.

Jackiewicz, I., & Jackiewicz, A. M. (2017). Benchmarking: teoria i praktyka. *Przedsiębiorczość i Zarządzanie, 18*(2, cz. 2 Zarządzanie humanistyczne i publiczne), 117–131.

Jeston, J., & Nelis, J. (2014). *Business Process Management*. London and New York: Routledge.

Jones, D., & Womack, J. (2011). *Seeing the Whole Value Stream.* Brookline: Lean Enterprise Institute.

Jurczuk A. (2018). Identification and Tracking of Process Inconsistencies in Manufacturing Enterprises. In S. Sahin (Ed.), *8th International Conference on Engineering, Project, and Product Management (EPPM 2017).* Lecture Notes in Mechanical Engineering (pp. 129–138). Cham: Springer.

Kanawaty, G. (1992). *Introduction to Work Study*. Geneva: ILO.

Kaplan, R. S., & Cooper, R. (1998). *Cost & Effect: Using Integrated Cost Systems to Drive Profitability and Performance.* Boston: Harvard Business School Press.

Kasse, T. (2008). *Practical Insight into CMMI®*. Norwood: ARTECH HOUSE.

Kauf, S. (2015). Orientacja procesowa w zarządzaniu publicznym łańcuchem dostaw. *Studia Ekonomiczne, 251,* 122–131.

Kaydos, W. (1999). *Operational Performance Management*. Boca Raton: CRC Press LLC.

Kerzner, H. (2017). *Project Management: A Systems Approach to Planning, Scheduling, and Controlling*. Hoboiken: Wiley.

Kobayashi, I. (1995). *20 Keys to Workplace Improvement*. Cambridge, MA: Productivity Press.

Kosieradzka, A. (2010). Applying Capability Maturity Model CMMI® to Production Management. In M. Fertsch (Ed.), *Innovative and Intelligent*

Manufacturing Systems (Monograph, Chapter 8, pp. 115–130). Poznań: Publishing House of Poznan University of Technology.

Kosieradzka, A. (2012). *Zarządzanie produktywnością w przedsiębiorstwie* [Productivity Management in the Entrprise]. Warszawa: C. H. Beck.

Kosieradzka, A. (2017). Maturity Model for Production Management. In K. Halicka & L. Nazarko (Eds.), *Procedia Engineering* (Vol. 182, pp. 342–349). Special Issue 7th International Conference on Engineering, Project, and Production Management. Elsevier. https://doi.org/10.1016/j.proeng.2017.03.109.

Kosieradzka, A., & Ciechańska, O. (2018). Impact of Enterprise Maturity on the Implementation of Six Sigma Concept. *MPER, 3,* 59–70. https://doi.org/10.21008/j.0239-9415.2017.074.01.

Kosieradzka, A., Kąkol, U., & Krupa, A. (2011). The Development of Production Management Concepts. *Foundation of Management, 3*(2), 55–74.

Kosieradzka, A., & Smagowicz, J. (2016). Analiza porównawcza modeli dojrzałości organizacji [Comparative Analysis of Organization Maturity Models]. In M. Ćwiklicki, M. Jabłoński, & S. Mazur (Eds.), *Współczesne koncepcje zarzadzania publicznego* (pp. 283–296). Kraków: Fundacja Gospodarki i Administracji Publicznej.

Kosieradzka, A., & Smagowicz, J. (2018). *Model dojrzałości organizacji w obszarze publicznego zarządzania kryzysowego* [Organisation Maturity Model in the Area of Public Crisis Management]. In *Zeszyty Naukowe Politechniki Śląskiej* (pp. 217–230). Seria: Organizacja i Zarządzanie z. 128.

Krogstie, J. (2016). Quality of Business Process Models. In *Quality in Business Process Modeling* (pp. 53–102). Cham: Springer.

Lee, R. G., & Dale, B. G. (1998). Business Process Management: A Review and Evaluation. *Business Process Management Journal, 4*(3), 214.

Liker, J. K. (Ed.). (1998). *Becoming Lean.* Portland: Productivity Inc.

Liker, J. K. (2004). *The Toyota Way. 14 Management Principles from the World's Greatest Manufacturer*. New York: McGraw-Hill.

Liker, J. K., & Meier, D. (2006). *Toyota Way Fieldbook.* New York: McGraw-Hill.

Linderman, K. K., Schroeder, R. G., & Sanders J. (2010, November). A Knowledge Framework Underlying Process Management. *Decision Sciences, 41*(4), 689–719.

Locher, D. (2011). *Lean Office and Service Simplified*. New York: CRC Press.

Looy, A. (2014). *Business Process Maturity: A Comparative Study on a Sample of Business Process Maturity Model*. Berlin, Heidelberg: Springer International Publishing.

Lunau, S. (Ed.). (2008). *Six Sigma + Lean Toolset*. Berlin, Heidelberg: Springer-Verlag.

Malinova, M., & Mendeling, J. (2018). Identifying do's and Don'ts Using the Integrated Business Process Management Framework. *Business Process Management Journal, 24*(4), 882–899.

Matla, J. (2011). *Dobre praktyki zidentyfikowane podczas samooceny w ramach projektu systemowego*. Warszawa: Ministerstwo Administracji i Cyfryzacji.

McCormack, K. (2001). Business Process Orientation: Do You Have It? *Quality Progress, 34*(1), 51–58.

Melnik, S. A., & Denzler, D. R. (1996). *Operations Management: A Value-Driven Approach*. Chicago: Irwin.

Mendling, J. (2012). Managing Structural and Textual Quality of Business Process Models. In *International Symposium on Data-Driven Process Discovery and Analysis* (pp. 100–111). Berlin, Heidelberg: Springer.

Mendling, J., Reijers, H. A., & van der Aalst, W. M. (2010). Seven Process Modeling Guidelines (7PMG). *Information and Software Technology, 52*(2), 127–136.

Molenda, M. (2016). Zasady ustalania mierników procesów w systemie zarządzania jakością. *Systemy Wspomagania w Inżynierii Produkcji, nr, 2*(14), 249–257.

Nadarajah, D., Syed, A. & Kadir, S. L. (2016). Measuring Business Process Management Using Business Process Orientation and Process Improvement Initiatives. *Business Process Management Journal, 22*(6), 1069–1078.

Nadler, G. (1967). *Work System Design the Ideal's Concept*. Homewood: Irwin.

Nakajima, S. (1988). *Introduction to TPM*. Portland: Productivity Press.

Netjes, M., Reijers, H., & van der Aalst, M. (2006). Supporting the BPM Life-Cycle with FileNet. In *Proceedings of the Eleventh CAiSE/IFIP8.1 International Workshop on Evaluation of Modeling Methods in Systems Analysis and Design* (pp. 497–508).

Nurcan, S., Schmidt, R., Bider, I., & Regev, G. (2012). Special Issue on Business Process Lifecycle: Design, Deployment, Operation and Evaluation. *Journal of Software: Evolution and Process, 24*(3), 231–236.

Oakland, J. S. (2014). *Total Quality Management and Operational Excellence*. New York: Routledge.

Oakland, J. S., & Oakland, R. J. (2019). *Statistical Process Control*. London and New York: Routledge.

OGC. (2010). *Management of Risk: Guidance for Practitioners*. London: TSO.

OMG. (2008). *Business Process Maturity Model (BPMM)*. https://www.omg.org/spec/BPMM/1.0/PDF.

P3O. (2008). *Portfolio, Programme and Project Offices*. Norwich: Office of Governmental Commerce, TSO.

Pande, P. S., Neuman, R. P., & Cavanagh, R. R. (2000). *The Six Sigma Way: How GE, Motorola, and Other Top Companies Are Honing Their Performance*. New York: McGraw-Hill.

Papazoglou, M. P., & Van Den Heuvel, W. J. (2007). Business Process Development Life Cycle Methodology. *Communications of the ACM, 50*(10), 79–85.

Pihir, I., Zajdela, N., & Dušak, V. (2010). Survey of Simulation Capabilities of the IBM WebSphere Business Modeler Business Process Modeling Tool on the Example of Processing a Loan Application. In *Proceedings of the International Conference on Information Technology Interfaces* (pp. 645–650). ITI.

PMI. (2003). *Organizational Project Management Maturity Model (OPM3)*. Newton Square: Project Management Institute.

Porter, M. E. (1985). *Competitive Advantage: Creating and Sustaining Superior Performance*. New York: Free Press.

Rosemann, M. (2018). The NESTT: Rapid Process Redesign at Queensland University of Technology. In *Business Process Management Cases* (pp. 169–185). Cham: Springer.

Rosemann, M., & vom Brocke, J. (2015). The Six Core Elements of Business Process Management. In J. vom Brocke & M. Rosemann (Eds.), *Handbook of Business Process Management 1* (pp. 107–122). Berlin, Heidelberg: Springer.

Rostek, K. (2010). Business Intelligence for SME. In E. Lechman (Ed.). *SMEs and Entrepreneurship* (Vol. II, pp. 164–190). Gdańsk: Gdańsk University of Technology Publishing House.

Rostek, K., Wiśniewski, M., & Gołuch-Trojanek, K. (2018). Uwarunkowania procesu wdrażania zintegrowanych systemów informatycznych w instytucjach publicznych. *Nowoczesne Systemy Zarządzania, 13*(4), 117–129.

Rother, M., & Harris, R. (2001). *Creating Continuous Flow*. Brookline: Lean Enterprise Institute.

Rother, M., & Shoock, J. (2009). *Learning to See: Value Stream Mapping to Create Value and Eliminate Muda*. Brookline: Lean Enterprise Institute.

Rummler, G. A., & Brache, A. P. (1995). *Improving Performance*. San Francisco: Jossey-Bass Publishers.

Scheruhn, H. J., von Rosing, M., & Fallon, R. L. (2015). Information Modeling and Process Modeling. In M. Von Rosing, H. Von Scheel, & A.

W. Scheer (Eds.), *The Complete Business Process Handbook: Body of Knowledge from Process Modeling to BPM, Volume 1* (pp. 511–550). Waltham, MA: Morgan Kaufmann.

Schwab, K. (2016). *The Fourth Industrial Revolution.* Cologny, Switzerland: World Economic Forum.

Senge, P. (2006). *The Fifth Discipline.* New York: Currency Doubleday.

Shankar, R. (2009a). *Process Improvement Using Six Sigma—A DMAIC Guide, Manufacturing Engineering.* Milwaukee, WI: ASQ Quality Press.

Shankar, R. (2009b). *Process Improvement Using Six Sigma—A DMAIC Guide.* Milwaukee: ASQ Quality Press.

Smagowicz, J. (2019). *Model dojrzałości zarządzania ryzykiem w publicznym zarządzaniu kryzysowym, rozprawa doktorska* [Risk Management Maturity Model in Public Crisis Management] (Doctoral Dissertation). Warsaw University of Technology, Faculty of Management, Warsaw.

Snopko, J. (2011). Kulturowe uwarunkowania dokonywania zmian organizacyjnych w urzędach gmin. *Zarządzanie Publiczne, 14*(2), 27–39.

Sujova, A., Simanova, L., & Marcinekova, K. (2019). Reengineering of Production Processes and Its Impact on the Financial Situation and Business Performance of the Company. *Engineering Management in Production and Services, 11*(3), 106–116. https://doi.org/10.2478/emj-2019-0025.

Syed, R., Bandara, W., French, E., & Stewart, G. (2018). Getting It Right! Critical Success Factors of BPM in the Public Sector: A Systematic Literature Review. *Australasian Journal of Information Systems, 22,* 1–39.

Szumowski, W. (2009). Zastosowanie podejścia procesowego w urzędach administracji samorządowej. *Prace Naukowe Uniwersytetu Ekonomicznego we Wrocławiu,* no. 52 Podejście procesowe w organizacjach, pp. 106–118.

Thome, B. (Ed.). (1993). *Principles and Practice of Computer-Based Systems Engineering.* Wiley Series in Software Based Systems. Chichester: Wiley.

Tregear, R., & Jenkins, T. (2007). *Government Process Management: A Review of Key Differences Between the Public and Private Sectors and Their Influence on the Achievement of Public Sector Process Management.*

Van der Aalst, W. (2013). *Business Process Management: A Comprehensive Survey.* In International Scholarly Research Notices, Volume 2013, Hindawi. https://doi.org/10.1155/2013/507984.

Virtual Corporation. (2020). *Business Continuity Maturity Model.* https://www.virtual-corp.com/bcmm.

Vom Brocke, J., & Sinnl, T. (2011). Culture in Business Process Management: A Literature Review. *Business Process Management Journal, 17*(2), 357–378. https://doi.org/10.1108/14637151111122383.

Von Rosing, M., Hove, M., von Scheel, H., & Morrison, R. (2015). BPM Center of Excellence. In M. von Rosing, H. von Scheel, & A. W. Scheer (Eds.), *The Complete Business Process Handbook: Body of Knowledge from Process Modeling to BPM* (Vol. 1, pp. 219–243). Waltham, MA: Morgan Kaufmann.

Weske, M. (2012). *Business Process Management: Concepts, Languages, Architectures*. Berlin: Springer.

Womack, J. P., & Jones, D. T. (1996). *Lean Thinking: Banish Waste and Create Wealth in Your Corporation*. New York: Simon & Schuster.

Womack, J. P., Jones, D. T., & Ross, D. (1990). *The Machine That Changed the World*. New York: Macmillan.

Woźniak, J., & Fester, A. (2015). Architektura korporacyjna–źródło korzyści dla przedsiębiorstwa. *Edukacja Ekonomistów i Menedżerów, 38*(4), 201–227.

Wróbel, Ł. (2017). *Zarządzanie zmianą i procesami biznesowymi w polskich przedsiębiorstwach*. Raport z badania. Webcon, Computerworld.

Ziemba, E., & Obłąk, I. (2014). Informatyczne wsparcie procesów w administracji publicznej. *Roczniki Kolegium Analiz Ekonomicznych/Szkoła Główna Handlowa, 33,* 619–646.

Zur Muehlen, M., & Ho, D. T. (2006). Risk Management in the BPM Lifecycle. In C. Bussler & A. Haller (Eds.), *Business Process Management Workshops* (pp. 454–466). Berlin, Heidelberg: Springer.

Index

A. Kosieradzka and K. Rostek, *Process Management and Organizational Process Maturity*,
https://doi.org/10.1007/978-3-030-66800-6

Q

R

The manufacturer's authorised representative in the EU is Springer
Nature Customer Service Centre GmbH, Europaplatz 3, 69115 Heidelberg,
Germany. If you have any concerns regarding our products, please
contact ProductSafety@springernature.com

Printed and bound by CPI Group (UK) Ltd, Croydon, CR0 4YY

30/09/2026

02225457-0007